A.E.G. Aircraft of WWI

A Centennial Perspective on Great War Airplanes

Jack Herris

Great War Aviation Centennial Series #16

This book is dedicated to all the WWI aviation enthusiasts who participate in or support the Dawn Patrol Rendezvous events at the National Museum of the USAF.

Acknowledgements

My sincere thanks to Greg VanWyngarden, Colin Owers, Bill Toohey, Tobias Weber, Dave Hooper, Dave Watts, and Richard Andrews for photographs, Bob Pearson for his color profiles, Aaron Weaver for photographs and image editing, Marty Digmayer and Colin Owers for scale drawings, and the Museum of Flight in Seattle for photographs. Thanks also to Steve Anderson for the original cover painting specifically for this book and Aaron Weaver for the cover design. I also want to thank Reinhard Zankl for his helpful information regarding production orders and serial numbers and for photographs. The late Peter M. Grosz provided a draft manuscript that was of significant help when writing this book. Any errors are my responsibility.

AEG G.IVk scale drawing courtesy of Ilya Sobolev of Copper State Models (**www.copperstatemodels.com**).

Color aircraft profiles © Bob Pearson. Purchase his CD of WWI aircraft profiles for $50 US/Canadian, 40 €, or £30, airmail postage included, via Paypal to Bob at: **bpearson@kaien.net**

Cover painting by Steve Anderson. Please see Steve's website at: **www.anderson-art.com**

For our aviation books in print and electronic format, please see our website at: **www.aeronautbooks.com**.
I am looking for photographs of the less well-known German aircraft of WWI. For questions or to help with photographs you may contact me at **jherris@me.com**

ISBN: **978-1-935881-28-5**

Design and layout: Jack Herris
Cover design: Aaron Weaver
Digital photo editing: Aaron Weaver & Jack Herris

Aeronaut Books

www.aeronautbooks.com

Table of Contents

Introduction	3
AEG 1914 Flying Boat	6
AEG S1	7
AEG Reconnaissance Two-Seaters	8
AEG B	8
AEG B.I	10
AEG B.II	11
AEG B.III	14
AEG C.I	15
AEG C.II	18
AEG C.III	20
AEG C.IV	23
AEG C.V	43
AEG C.VI	44
AEG C.VII	44
AEG C.VIII	46
AEG Fighters	47
AEG D.I	47
AEG Dr.I	50
AEG Armored Aircraft	52
AEG J.I	52
AEG J.II	66
AEG J.III	82
AEG DJ.I Triplane	83
AEG DJ.I Biplane	84
AEG G.IVk	86
AEG Bombers	91
AEG K.I	92
AEG G.I	93
AEG G.II	94
AEG G.III	109
AEG G.IV	121
AEG G.V	137
AEG N.I	150
AEG R.I & R.II	160
AEG Miscellaneous Aircraft	162
AEG-Wagner Monoplane	162
AEG Igel Project	162
AEG Naval R-Planes	162
AEG in Retrospect	163
Bibliography	164
Index	164
Scale Drawings	
AEG J.I (1/48 – MD)	165
AEG J.II (1/48 – MD)	168
AEG N.I (1/48 – MD)	171
AEG D.I (1/48 – CAO)	174
AEG G.II (1/48 – CAO)	176
AEG G.III (1/48 – CAO)	178
AEG G.IVk (1/72 – MD)	180
AEG G.V (1/72 – MD)	183
Afterword	186

Forward

AEG was one of the most important German aircraft manufacturers of WWI and it is a pleasure to be able to document their aircraft. Perhaps most well-known for their twin-engine bombers, AEG produced significant two-seaters, including the armored J.I and J.II that were important combat aircraft yet seldom documented or recognized. The little-known AEG G.V that reached operations late in the war may have been the best G-type to see combat. The AEG D.I of 1917 was the fastest German fighter until very late in 1918, but production was cancelled due to unexplained fatal accidents, perhaps robbing the German air service of the superior fighter it desperately needed in late 1917. Finally, the AEG G.IVk anti-tank aircraft that reached the front at the end of the war previewed the anti-tank helicopters of today. I hope you enjoy this tribute to AEG and their aircraft.

Jack Herris

Above: The business end of an AEG G.IV carrying bombs.

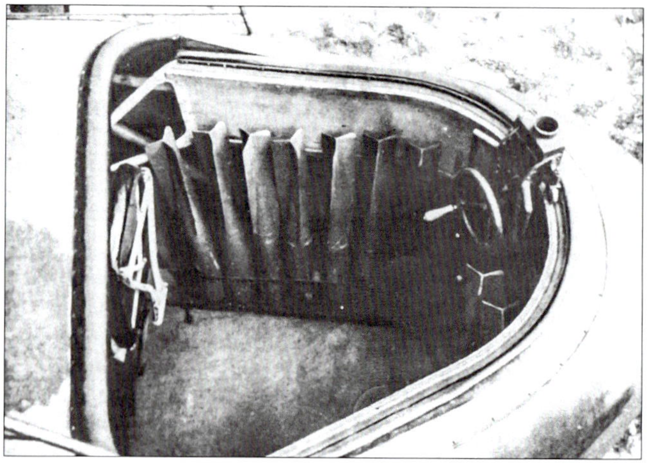

Above: The bombardier/front gunner's cockpit of an AEG G.IV; up to 18 12.5 kg P.u.W. bombs could be stored in the cockpit for the bombardier to drop by hand. The crewman's folding seat is at left and the control wheel for adjusting the gun mount is at right.

AEG Introduction

Above: AEG built a wide variety of warplanes but is probably best known for its twin-engine bombers. Above is G.V G.625/18, an example of the final production AEG bomber. The G.V was based on the G.IV with its wing enlarged in span and area for more lift to carry the desired 1,000 kg bomb load, requiring an additional bay of struts. A 'box' tail gave better control during asymmetric thrust after loss of an engine, and the Flettner tabs on the ailerons reduced control forces, making the large bomber easier for the pilot to fly. The G.V was powered by two 260 hp Mercedes D.IVa engines.

The Allgemeine Elektrizitäts-Gesellschaft (general electric company – AEG) was one of Germany's major high-tech industrial companies, and characteristically was one of the first to establish an aviation department. In 1910 AEG set up their new department in Hennigsdorf, where it started by building an all-wood biplane based on the Wright Flyer that AEG had purchased. This was followed by another aircraft that used the original Wright wings and a *Taube* with steel-tube fuselage and wooden wings. After the Taube crashed, AEG noted that the wings were totally destroyed while the steel-tube fuselage remained intact. As a result, AEG decided to specialize in aircraft of steel-tube construction and became the first German aircraft manufacturer to build metal-framed aircraft.

In November 1911 the AEG Z2 biplane was built completely of steel tube with the exception of wood wing ribs. In summer 1912 AEG built its Z3 biplane with foldable wings to make transportation by rail or truck easier. Also in 1912 AEG moved its airfield from Teltow to Nieder-Neuendorf to be closer to Hennigsdorf. In 1913 a flying school for army officers was opened, and AEG built its Z4, Z5,

Above: Parachute attached to an AEG G.IV of *KG* 4. Parachutes were normally issued to German fighters pilots starting in June 1918 so this may have been a test setup.

and Z6 biplanes with various engines. These biplane trainers were used at the school, where AEG trained 320 pilots between the school's opening in 1913 and spring 1917, when the school had to be closed due to greater priority being given to test flying and acceptance testing.

AEG increased its aircraft production by sourcing parts and sub-assemblies from their Berlin turbine plant and their Brunnenstrasse factory, and new assembly shops in Nieder-Neuendorf were opened in

Frontbestand Inventory of All AEG Aircraft at the Front

Manufacturer and Type		1914 31 Aug	1914 31 Oct	1914 31 Dec	1915 28 Feb	1915 30 Apr	1915 30 Jun	1915 31 Aug	1915 31 Oct	1915 31 Dec	1916 28 Feb	1916 30 Apr	1916 30 Jun	1916 31 Aug	1916 31 Oct	1916 31 Dec	1917 28 Feb	1917 30 Apr	1917 30 Jun	1917 31 Aug	1917 31 Oct	1917 31 Dec	1918 28 Feb	1918 30 Apr	1918 30 Jun	1918 31 Aug
AEG	B.I	3	2	3	8	1	2		1	2								1								
	B.II					8	8	9																		
AEG	C.I						4	10	28	26	34	34		1	1	1	1	1								
	C.II											1														
	C.IV								1						9	42	90	124	127	63	88	68	40	26	29	34
AEG	D.I																			1						
AEG	G.I							1	5			5														
	G.II						2	5	10	13	12	2	4	4	4	2	2	1	1							
	G.III												6	16	22	21	22	9								
	G.IV									1								5	9	15	15	35	37	54	74	51
	G.																	3								
AEG	J.I																		4	1	11	35	55	66	43	28
	J.II																							15	65	63
	J.																		3							
AEG	N.I																				2	31	37	19	7	4

AEG C-Type Production Orders

Serial Numbers	Qty	Order Date & Notes
AEG C.I (70 Total)		
C.74–79/15	6	Unknown
C.82–93/15	12	April 1915
C.130–153/15	24	May 1915
C.324–348/15	25	May 1915
C.1366–1368/17	3	Unknown
AEG C.II (37 Total)		
C.869–888/15	20	September 1915
C.1887–1902/15	16	October 1915
C.9421/16	1	C.II prototype?
AEG C.IV (495 Total)		
C.1024–1123/16	100	March 1916
C.6575–6674/16	100	October 1916 (note 1)
C.1700–1799/17	100	April 1917
C.4800–4899/17	100	(May) 1917
C.7052–7126/17	75	July 1917 (note 2)
?	8	December 1917 (note 3)
C.1100–1111/18	12	February 1918 (note 3)
AEG C.IV(Fok) (200 Total)		
C.240–439/17	200	January 1917
AEG C.IVa(Fok) (100 Total)		
C.6500–6699/17	100	January 1917 (note 4)

Notes:
1. Initially order was for 100 Rumpler C.IV(AEG)
2. 25 of these for Turkey, 50 for Germany
3. Special equipment for tropics
4. 200 ordered but only 100 built

AEG G-Type Production Orders

Serial Numbers	Qty	Order Date & Notes
AEG G.II (27–28 Total)		
G.2–7/15	6	April 1, 1915
G.19–30/15	12	May 6, 1915
G.46–51/15	6	September 7, 1915
Unknown	(3–4)	Sept. 22, 1915 (note 1)
AEG G.III (46 Total)		
G.8/15	1	April 1, 1915 (note 2)
G.52–56/15	5	Sep. 7, 1915 (note 3)
G.210–239/15	30	Dec. 1, 1915 (note 4)
G.143–152/16	10	March 1916 (note 4)
AEG G.IV (320 Total)		
G.153–192/16	40	March 1916 (note 5)
G.1095–1144/16	50	December 1916
G.560–609/17	50	September 1917
G.844–893/17	50	December 1917
G.545–619/18	75	April 1918
G.1215–1264/18	50	July 1918
AEG G.IVk (5 Total)		
G.500–504/18	5	March 1918
AEG G.V (50 Total)		
G.620–644/18	25	April 1918
G.1700–1724/18	25	October 1918

Notes:
1. 12 ordered but only 3–4 delivered
2. Prototype, single rudder, serial unconfirmed.
3. G.53/15 had triple rudders
4. Single rudder
5. Includes some AEG G.IVb
6. The AEG G.I serial was G.1/15.

AEG J-Type Production Orders

Serial Numbers	Qty	Type & Order Date
J.150–159/17	10	AEG J.I; March 1917
J.200–249/17	50	AEG J.I; May 1917
J.428–527/17	100	AEG J.I & J.II; July 1917
J.175–374/18	200	AEG J.II;March 1918
J.375–574/18	200	AEG J.II (see notes below)
J.575/18	1	AEG J.III Prototype?

Many thanks to Reinhard Zankl for this data.

AEG D.I Production Orders

Serial Numbers	Qty	Order Date
D.4400–4402/17	3	June 1917
D.5500–5019/17	20	July 1917

AEG N.I (C.IVn) Production Orders

Serial Numbers	Qty	Order Date
C.9321–9420/16	100	December 1916
N.110–209/17	100	November 1917

Note: Some or all of the first production order were given the N designation

July 1915 to become the AEG Flugzeugfabrik.

AEG followed these early biplanes with a series of B-type and related C-types, many of which were produced in quantity. In particular the AEG C.IV was an important design; many C.IV aircraft served at the front and it served as the basis for the AEG N.I single-engine night bomber and the AEG J.I and J.II armored infantry aircraft that became increasingly important in combat.

AEG also built a series of two-engine aircraft that were developed into the most efficient German bombers of the war. AEG did build one unsuccessful Giant bomber and were working on more advanced giant bombers for the Army and Navy when the war ended. AEG was not successful building single-seat fighters; none of the prototypes built were ordered into production despite the speed of the AEG D.I.

The National Automobil AG, or NAG, was an AEG subsidiary that built 765 Conrad C.III(Nag) six-cylinder engines for trainers during the war. Based on its experience building turbines, AEG became a leader in supercharger development, although none went into production during the war.

AEG employed a peak of 3,162 employees in May 1918, growing from 224 workers on January 1, 1915. The number of aircraft AEG built during the war depends on the source consulted, ranging from 1,790 stated in the postwar Inter-Allied report to 1,887 based on a count of serial numbers. The actual total is likely between these extremes but is not known precisely. AEG production orders by major type are summarized on page 5.

AEG 1914 Flying Boat

Above: The AEG monoplane flying boat completed in July 1914. Only one was built; the Navy evaluated the aircraft and judged that it was unsuitable for naval service. The rod mounted above the engine was for lifting the boat out of the water.

AEG completed a monoplane flying boat with side-by-side seating and powered by a 150 hp Benz Bz.III pusher engine in July 1914, just before the war started. Intended for the *Ostseeflug Warnemünde* competition, the AEG flying boat was impressed by the Navy when Germany mobilized on 1 August 1914. There is no record of tests in the naval war diary, but it is possible some flights were performed by AEG personnel who had come to Warnemünde for the *Ostseeflug* and were required to remain until released by the Navy. The Navy judged the AEG flying boat as unsuitable for military duties and it was returned to the factory August 26, 1914.

AEG 1914 Flying Boat Specifications

Engine:	150 hp Benz Bz.III	
Wing:	Span	16.00 m
General:	Length	10.00 m
	Height	3.65 m

AEG S1

In early 1914 AEG built a floatplane variant of the Z6 landplane for the German Navy. The AEG S1 had folding wings for shipboard use. Completed in May, the S1 (work number 42, Marine Number 45) was powered by a 150 hp Benz Bz.III. It was delivered to the naval seaplane command at Holtenau in June for testing. The S1 was returned to AEG for repair after several flights and was returned to Warnemünde on August 21, 1914. Tests flown in September reveals that the S1 was not acceptable for service; carrying only the pilot and 20 liters of fuel, the S1 lost speed in a turn to such an extent that it became dangerously unstable. On October 27 the S1 was again returned to AEG, this time for conversion to a landplane. Navy landplane number LF35 had been reserved for the converted S1 but this number was re-assigned to the Aviatik WP18 in October.

AEG entered a second AEG S1 (w/n 43) in the *Ostseeflug Warnemünde* competition supported by the *National Flug Spende* and German Navy that was scheduled for August 1–10, 1914. With the general mobilization on August 1, the 22 *Ostseeflug* competitors were impressed and held at Warnemünde for acceptance testing. On August 11, naval pilots reported that the heavy S1 possessed insufficient climb. Fitted with lighter floats, additional trials were conducted by AEG pilots at Warnemünde between August 19 and October 3 but showed no improvement. Consequently, on October 27 the S1 was shipped to Hennigsdorf for conversion to a landplane.

AEG S1 Specifications

Engine:	150 hp Benz Bz.III	
Wing:	Span, Upper	16.00 m
	Chord, Upper	2.00 m
	Gap	2.00 m
General:	Length	11.80 m
	Height	3.80 m

Above: The unsuccessful AEG S1 floatplane being prepared for flight tests in May 1914 on the Havel See near the Hennigsdorf factory. The wings could be folded back for storage. Despite being powered by a 150 hp Benz Bz.III engine, the S1 was too heavy and had poor climb; it was not accepted by the Navy.

AEG Reconnaissance Two-Seaters

AEG produced a variety of two-seaters. The B-type reconnaissance aircraft were purchased in small numbers, and it was not until the C.I that an AEG two-seater was built in any quantity. The next type to be built in any quantity was the C.IV. The AEG C.IV was significant for several reasons. First, it was built and used in far greater numbers than any preceding AEG aircraft, and proved successful at the front. Next, the basic C.IV design was used for the N.I two-seat night bomber and the armored J.I infantry cooperation aircraft. In turn the J.I was developed into the J.II, the most numerous J-type built. Later AEG C-types remained prototypes only.

AEG B-Types

After some unspecified problems were overcome the air service ordered six B-type (unarmed, two-seat biplanes) of the Z6 type for training and operations that were delivered in early 1914.

Above: AEG Z1 was a pre-war, unarmed two-seater.

Below: Early AEG B-type; the nose wheel was to prevent nose-overs on landing.

Above: Completed AEG ZIII and Z6 biplanes at the factory pre-war.

Below: AEG B-type trainers at AEG Flugplatz Nieder-Neuendorf. These aircraft were owned by the flying school and the numbers were their school numbers, not military serial numbers.

AEG B.I

The AEG Z6 gave satisfactory service and the air service ordered 20 improved aircraft (Z6a) in April 1914. These were powered by a variety of 100–120 hp NAG, Mercedes, or Benz six-cylinder engines. A small number of these aircraft served at the front between August 1914 and December 1915, their steel-tube structures proving their reliability and durability. When *Idflieg* published the new aircraft designations in August 1915 the AEG Z6a biplanes were retroactively designated AEG B.I even though they were no longer in production.

Above: AEG B.I, the retroactive designation for the AEG Z6a. (The Peter M. Bowers Collection/The Museum of Flight)

Left: AEG B.I B.274, perhaps after a hard landing. (The Peter M. Bowers Collection/The Museum of Flight)

AEG B.II

In late 1914 *Idflieg* ordered the improved AEG B.II powered by the 150 hp Benz Bz.III. This aircraft, company designation Z9, was a more compact development of the earlier B.I. Nothing is know about designs Z7 and Z8; these were probably unbuilt projects. The first production AEG B.II was rolled out in February 1915 and arrived at the front in April 1915. According to AEG records, ten aircraft were delivered and some served at the front until August–September 1915. Some B.II biplanes were re-engined with a 120 hp Mercedes D.II engine and fitted with dual controls for training purposes.

AEG B.II Specifications

Engine:	150 hp Benz Bz.III	
	120 hp Mercedes D.II	
Wing:	Span Upper	13.07 m
	Span Lower	11.83 m
	Chord Upper	1.60 m
	Chord Lower	1.60 m
	Gap	1.95 m
	Area	36.0 m^2
General:	Length	7.95 m
	Track	2.30 m
	Empty Weight	710 kg
	Loaded Weight	1,125 kg
Maximum Speed:		130 kmh
Climb:	1000m	4.5 min

Below: Front view of an AEG B.II. (The Peter M. Bowers Collection/The Museum of Flight)

Below: AEG B.II B.283 shows its engine and radiator details and the ability to fold its wings for storage. (The Peter M. Bowers Collection/The Museum of Flight)

Above: This view of AEG B.II B.283 shows the overall shape and many national insignia on the wings, but none on the fuselage. (The Peter M. Bowers Collection/The Museum of Flight)

Below: AEG B.II in flight. This aircraft has national insignia on the fuselage.

Above: AEG B.II B.244 in flight. The wings carry a surplus of national insignia but there are none on the fuselage.

Below: AEG B.II B.260/14 ready for flight. (The Peter M. Bowers Collection/The Museum of Flight)

Above: AEG B.II with wings folded for storage. The propeller also has a protective cloth covering.

AEG B.III

The AEG B.III (company designation Z10) was built in January 1915 as a lightened derivative of the B.II. Visually, the difference was a rounded fin and rudder and use of the 160 hp Mercedes D.III six-cylinder engine. As far as is known, only one B.III was built. Performance testing of the B.III compared to the B.II showed the B.III had a poorer climb despite being 10% lighter and having a slightly more powerful engine. Investigation revealed the bulged center-section gravity tank interfered with airflow from the propeller; when the tank was removed and the wing gap increased the B.III achieved the expected climb rate. As a result, starting with the AEG C.IV the center-section tank and airfoil radiator were eliminated from AEG designs. This improvement was later confirmed by wind-tunnel tests at Göttingen.

Below: The sole AEG B.III was powererd by a 160 hp Mercedes D.III.

AEG C.I

The AEG C.I (company designation KZ9 for *Kampf-Zweidecker* 9, or battle biplane 9) was derived from the earlier, unarmed B.II by adding a machine-gun turret; there was no fixed gun for the pilot. Although structurally the same as the B.II, with steel-tube construction except for wooden wing ribs, the C.I had a different wing planform than the B.II; however, the wing-folding mechanism was retained. *Idflieg* ordered six C.I aircraft in early 1915 and the C.I prototype was completed in March. Additional C.I aircraft were ordered, a total of about 70 being delivered.

The AEG C.I was among the first C-type aircraft to reach the front, serving from June 1915 to April 1916. Climb rate was mediocre despite the C.I being lighter than most contemporary C-types. Due to its sturdy metal airframe the C.I was used well into 1917 as a trainer after being removed from the front.

This Page: AEG C.I 90/15 (above) wears black and white bands on its fuselage and carries white identification pennants. Powered by a 150 hp Benz Bz.III like the AEG B.II, the AEG C.I still retained the side radiators normally associated with B-types. About 70 AEG C.I reconnaissance two-seaters were built. The C.I below may be 90/15 but that is not confirmed.

AEG C-Type Specifications					
	C.I	C.II	C.III	C.IV	C.IV(Fok)
Engine	150 hp Benz Bz.III	150 hp Benz Bz.III	150 hp Benz Bz.III	160 hp Mercedes D.III	160 hp Mercedes D.III
Span Upper	13.07 m	11.95 m	12.00 m	13.00 m	12.99 m
Span Lower	11.83 m	11.45 m	11.30 m	12.50 m	12.36 m
Chord Upper	1.60 m	1.85 m	1.60 m	1.65 m	1.65 m
Chord Lower	1.60 m	1.85 m	1.60 m	1.65 m	1.65 m
Gap	1.95 m	1.80 m	1.40 m	1.95 m	1.95 m
Wing Area	36.0 m^2	39.0 m^2	33.90 m^2	39.0 m^2	—
Length	7.95 m	7.09 m	6.50 m	7.20 m	7.22 m
Track	2.30 m	2.15 m	2.15 m	2.15 m	2.16 m
Empty Weight	710 kg	680 kg	687 kg	800 kg	901 kg
Loaded Weight	1,125 kg	1,200 kg	1,232 kg	1,320 kg	1,414 kg
Maximum Speed	130 kmh	138 kmh	145 kmh	158 kmh	—
Climb, 1000m	11 min.	8 min.	—	6 min.	4.8 min.
Climb, 2000m	25 min.	18 min.	—	12.5 min.	12.1 min.
Climb, 3000m	55 min.	41 min.	—	23 min.	21.9 min.
Climb, 4000m	—	—	—	38 min.	35.4 min.
Armament	1 flexible machine gun, small bombs	1 flexible machine gun, 4x10 kg bombs	1 flexible machine gun	1 fixed & 1 flexible machine gun	1 flexible machine gun

Above: AEG C.I 139/15 has broken its back in a hard landing. It wears national insignia on its fuselage as well as its rudder.

Facing Page: AEG C.I 75/15 was powered by a 150 hp Benz Bz.III and still used side radiators.

AEG C-Type Specifications

	C.V	C.VII	C.VIII	C.VIIIDr
Engine	220 hp Mercedes D.IV	160 hp Mercedes D.III	160 hp Mercedes D.III	160 hp Mercedes D.III
Span Upper	13.26 m	11.10 m	9.50 m	11.20 m
Span Lower	12.45 m	10.05 m	9.10 m	10.40 m
Chord Upper	1.75 m	1.55 m	1.74 m	1.45 m
Chord Lower	1.75 m	1.30 m	1.33 m	0.82 m
Gap	2.07 m	1.85 m	1.60 m	1.00 m
Wing Area	41.5 m^2	26.0 m^2	22.67 m^2	31.0 m^2
Length	7.60 m	6.20 m	6.20 m	6.90 m
Track	2.30 m	2.00 m	2.10 m	1.90 m
Empty Weight	900 kg	758 kg	800 kg	800 kg
Loaded Weight	1,432 kg	1,118 kg	1,160 kg	1,160 kg
Maximum Speed	165 kmh	175 kmh	170–190 kmh	158 kmh
Climb, 1000m	7 min.	4 min.	3.8 min.	3.8 min.
Climb, 2000m	13 min.	—	—	—
Climb, 3000m	22 min.	—	—	—
Climb, 4000m	37.5 min.	—	—	—
Armament	1 fixed & 1 flexible machine gun	1 fixed & 1 flexible machine gun	1 fixed & 1 flexible machine gun	1 fixed & 1 flexible machine gun
Note: C.VIII Triplane middle wing had 10.80 m span, 0.82 m chord. Climb to 5,000 m in 34 min.				

AEG C.II & C.IIa – C.IIe

The AEG C.II was basically a lightened C.I with a modified wing planform. Like the C.I it was powered by the 150 hp Benz Bz.III engine. It was 122 kg lighter than the C.I, giving it better performance, but airframe strength was compromised, resulting in a number of failures during load tests. Modifications to strengthen the airframe took time and the first six C.II aircraft were not delivered until April 1916. One C.II was sent to the front that month, but was soon removed from the front and sent to join the other C.II aircraft for training use.

From early 1915 through the spring of 1916, AEG modified C.II biplanes with different airfoil sections, chord, span, and gap, with the different configurations given factory (not *Idflieg*) designations C.IIa through C.IIe. Data was collected that led to aerodynamic improvements to wing and radiator design, some of which were key in the design of the AEG C.IV.

The AEG C.IIe tested an armored, annular radiator (Patent 299,725) in which the cooling air was forced through the radiator by a fan attached to the propeller. This particular development was driven by combat experience with the armored AEG J.I, which demonstrated that the radiator was the most vulnerable part of the armored aircraft during low-level ground attacks.

Above: AEG C.II 885/15 was from the first C.II production batch.

Left: AEG C.II, likely 885.15. The C.II retained the primitive side radiators used in the AEG B-types.

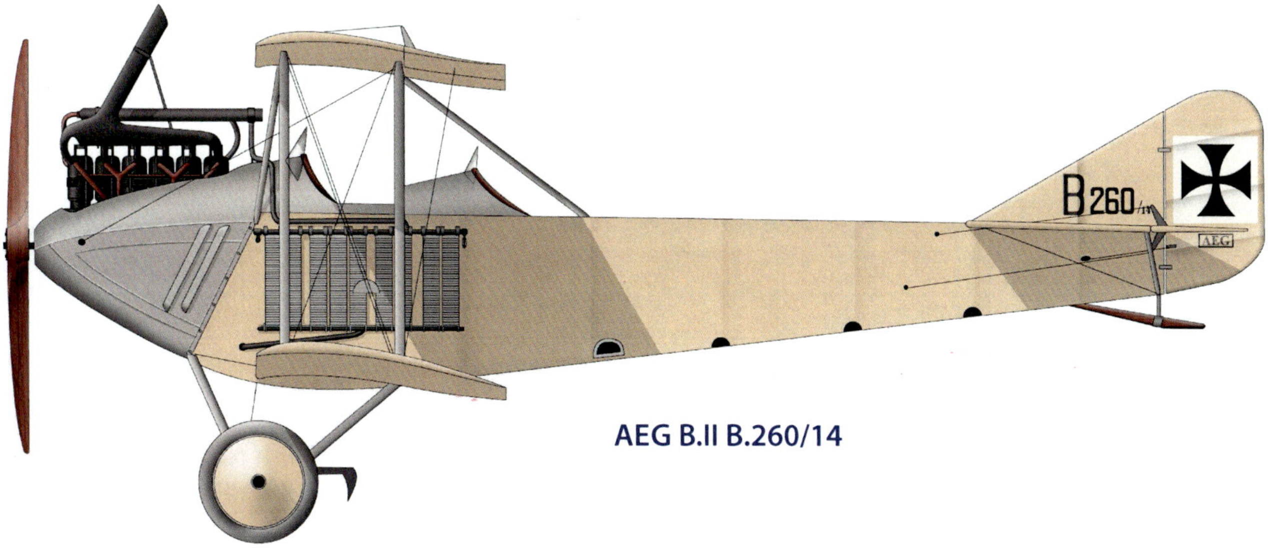

AEG B.II B.260/14

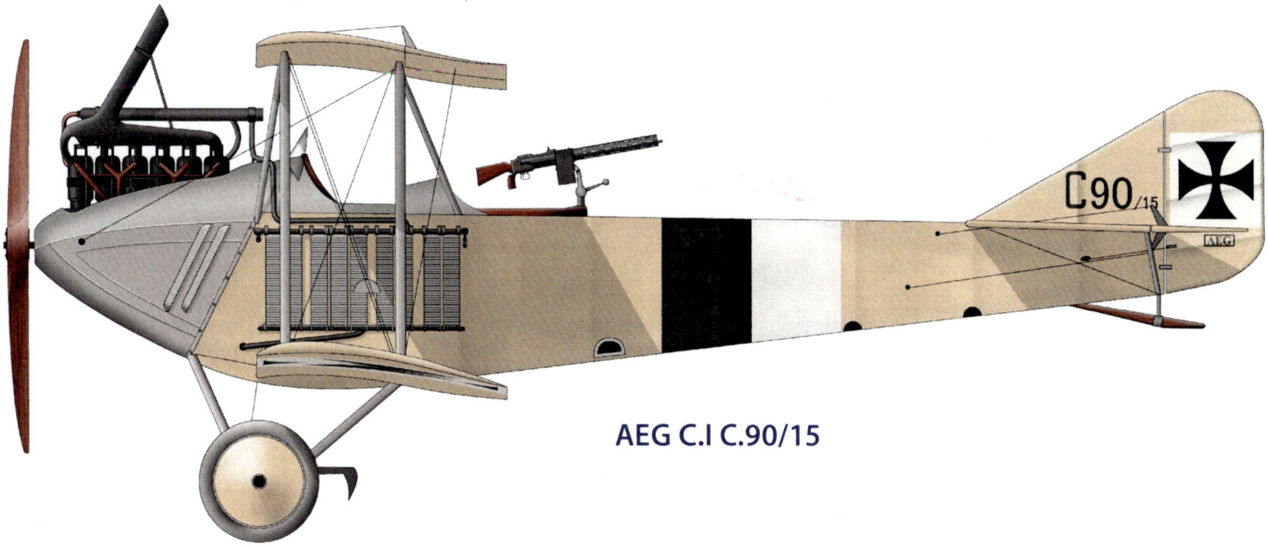

AEG C.I C.90/15

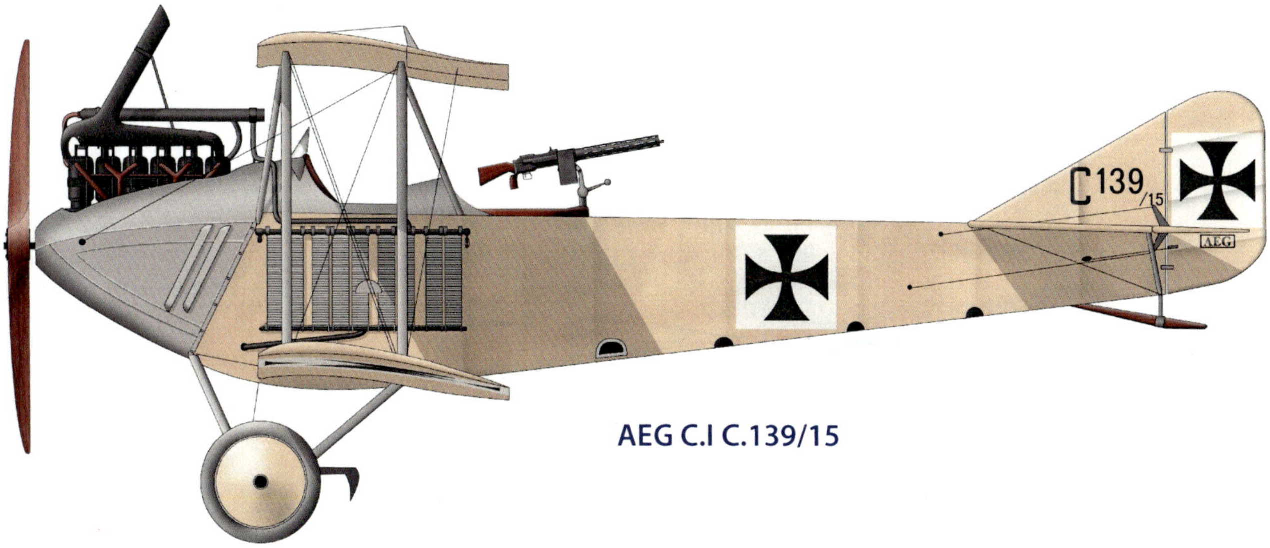

AEG C.I C.139/15

AEG C.IV in Color

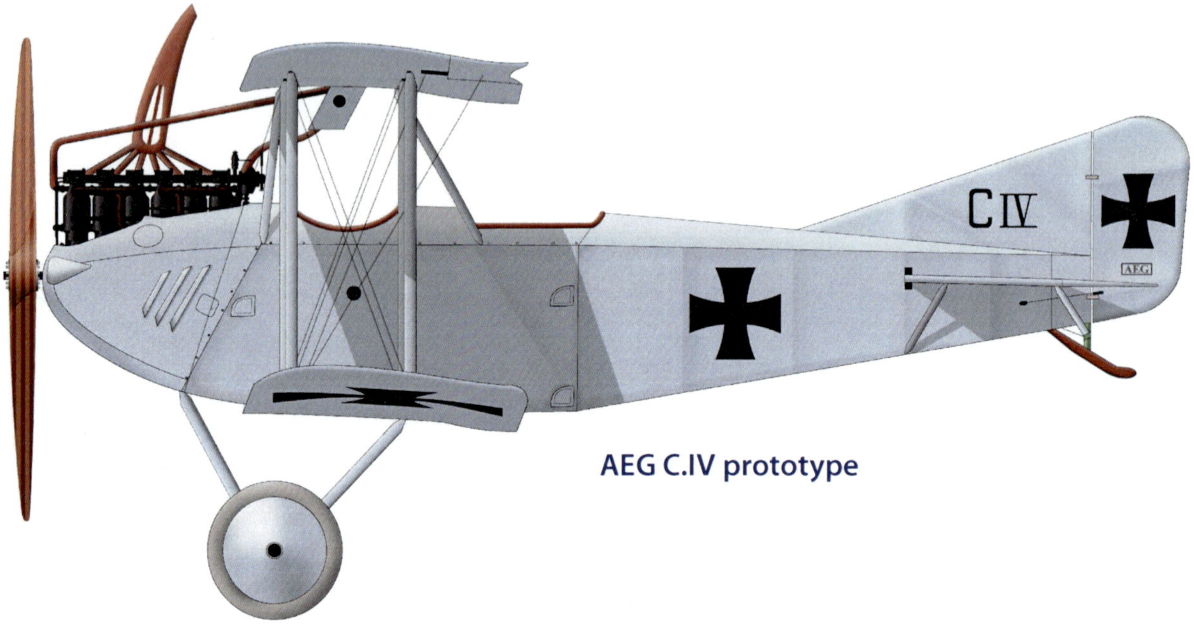

AEG C.IV prototype

AEG C.IV C.6674/16 of *Schusta* 27

AEG C.IV C.1748/17

AEG C.IV

Above: AEG C.IV 1715/17 is run up before its next mission while the aircrew and ground crew pose for a team photograph. The C.IV, powered by the 160 hp Mercedes D.III, was the first AEG C-type to have a synchronized gun for the pilot.

The AEG C.II had been insufficiently robust for front-line service and could only be used for training, and the C.III did not have the performance of the Roland C.II. To develop a successful C-type AEG based its C.IV design on the results of the engineering tests done on the C.IIa through C.IIe development prototypes. Attention was also given to speeding production through simplifying the design, rationalizing assembly methods, and building welding jigs to enable production by semi-skilled labor. Outwardly similar to the previous C.I and C.II, the C.IV embodied many detailed structural and aerodynamic improvements and a new wing design that eliminated the unnecessary wing-folding mechanism. Due to the unfortunate experience of the C.II, the C.IV airframe was strengthened and used structural joints machines from solid billets instead of built-up from welded sections. As a result the C.IV had improved strength and durability compared to any of its AEG predecessors.

Powered by a 160 hp Mercedes D.III engine, the C.IV prototype first flew in March 1916. *Idflieg* awarded AEG a production contract for 100 C.IV aircraft in March 1916 pending successful completion of the static load tests. Unfortunately, a wing spar failed load testing in June and *Idflieg* refused to accept the AEG C.IV. *Idflieg* then suggested that AEG build the all-wood Rumpler C.IV under license, but this was clearly not suited for the AEG production methods geared to metal-framed aircraft and the proposal was quickly dropped. In the meantime the AEG C.IV had managed to pass its load tests and was approved for service, the first aircraft reaching the front in October 1916.

At the front the AEG C.IV soon gained a reputation for reliability and speed, but was not an easy aircraft to fly compared to the well-liked Albatros C.III and DFW C.V and some units did not appreciate it. In addition to reconnaissance it also had a useful bombing ability. Never famous like the

Above: The AEG C.IV prototype (second version) benefitted from an improved steel-tube structure compared to the C.II.

shapely Roland C.II nor the exceptional Rumpler C.IV, the AEG C.IV was a solid, reliable aircraft that was capable enough to be developed into the production AEG N.I and AEG J.I and J.II.

Its robust metal airframe recommended it for service in the Middle-East where weather and temperature extremes were typical, and C.IV aircraft modified for the heat with larger radiators were delivered to both Turkish units and German units serving in the area. However, the aircraft's structure required welding equipment for even minor repairs, a factor that had been inexplicably overlooked when units using the type quickly found a nearly complete absence of the needed equipment in the area. This small but critical problem seriously reduced C.IV serviceability in the theater.

In January 1918 work was in progress to install a flexible 2cm Becker cannon in a C.IV modified with a larger observer's gun ring. The gun ring operated smoothly but the gunner's position was too cramped for proper loading and aiming of the cannon, and the project was cancelled in April.

AEG C.IV v.R.

AEG lengthened the C.IV fuselage to improve its flying and landing qualities; the modified aircraft was designated the AEG C.IV v.R. (*verlängerter Rumpf* – lengthened fuselage). This modification is not mentioned in any existing *Idflieg* records, although the AEG J.II was given a longer fuselage for the same reasons, as were some AEG J.I aircraft. To what extent the long fuselage was incorporated in late production aircraft is unknown.

AEG C.IV(Fok)

By the end of 1916 the Fokker company badly needed a suitable aircraft to manufacture because its own designs had been restricted to training service due to poor design and quality control. To maintain production output and the existing workforce, *Idflieg* contracted for license production of the AEG C.IV for training service. Fokker was the only other German company with the required welding skills for airframes, and in January 1917 received an order for 200 trainers as the AEG C.IV(Fok). Production deliveries began in July 1917 and the first production aircraft was type-tested in August 1917; deliveries were completed in December 1917.

AEG C.IVa(Fok)

Idflieg ordered 200 AEG C.IVa(Fok) trainers from Fokker in June 1917. The C.IVa(Fok) differed from the C.IV(Fok) in having a 180 hp Argus As.III engine in place of the 160 hp Mercedes D.III; the Mercedes was needed for Albatros and Pfalz fighter production and the Argus was readily available. Fokker delivered 100 of these aircraft between November 1917 and March 1918. However, the remaining 100 aircraft were cancelled to enable Fokker to use the production capacity for its new D.VII fighter.

Above: AEG C.IV prototype (second version) in April 1916 at the *Idflieg* hangar at Adlershof.
Below: AEG C.IV 1113/16 of the first production batch has suffered a landing accident.

Above: AEG C.IV 1103/16 from the first C.IV production batch was captured January 23, 1917 and put on display.

Below: AEG C.IV(Fok) airframes stored at the Fokker factory waiting for engines to be delivered. The streaked finish typical of Fokker F.I and Dr.I triplanes built in parallel with this production order clearly identifies these aircraft as built by Fokker.

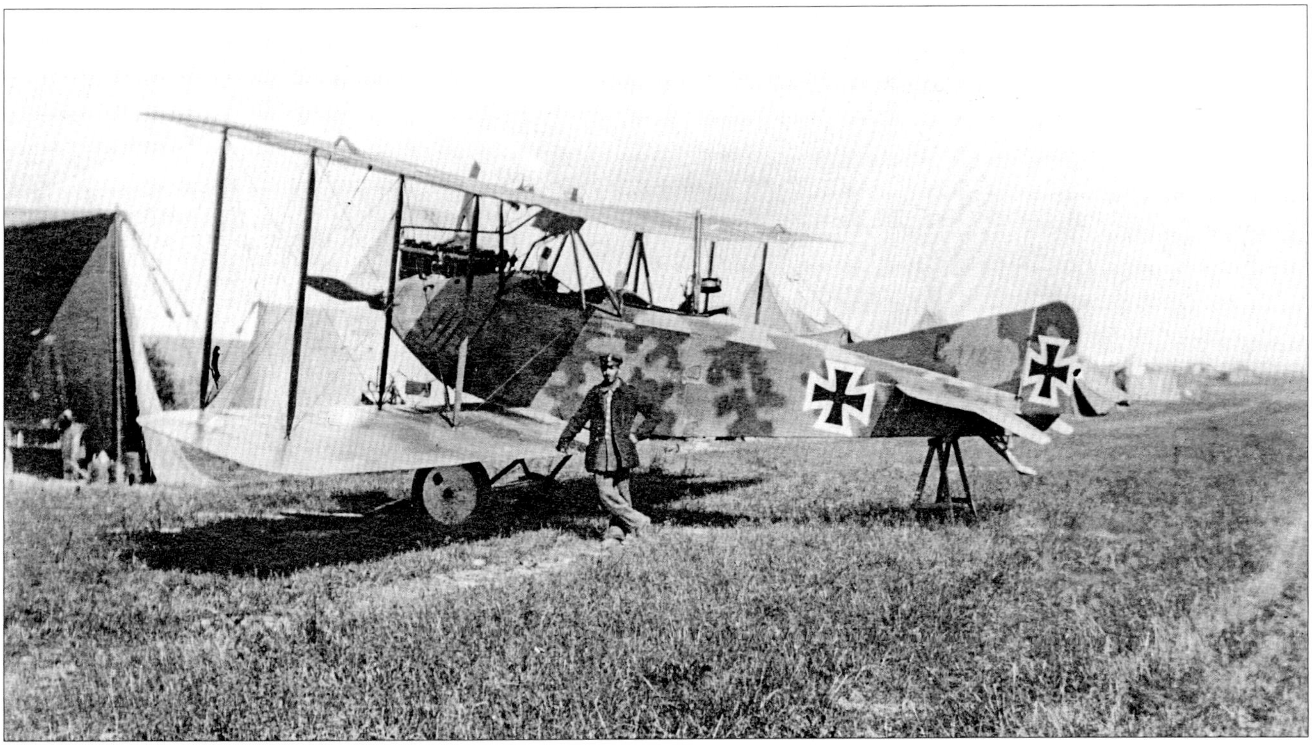

Above: A crewman strikes a jaunty pose by AEG C.IV 1748/17 of the third production batch. The two-color sprayed camouflage typical of AEG practice at the time shows up nicely.

Below: AEG C.IV 7123/17 v.R. (*verlängerter Rumpf* = lengthened fuselage) serving postwar with the Polish air force. Large serial numbers were used with training aircraft.

AEG C.IV in Color

AEG C.IV C.4871/17 of *FAA* 303 based in Palestine

AEG C.IV(Fok) trainer

AEG C.IV(Fok) trainer in postwar Netherlands service

Above: An AEG C.IV photographed in flight displays the lines typical of most AEG C-types.

Below: AEG C.IV 1724/17 after a landing accident. The serial number painted in large numbers on the fuselage is unusual.

AEG C.IV in Color

AEG C.IV C.6590/16

AEG C.IV in Turkish markings

AEG C.IVv.R C.7123/17 in postwar Polish service

Above: AEG C.IV 1762/17 of the third production batch.

Below: *Hptm*. Kieser in front of AEG C.IV 4871/17 serving with *Flieger Abteilung* 303 in Palestine.

Above: AEG C.IV 4886/17 of *Flieger Abteilung* 303 in Palestine being moved by unit personnel.

Below: AEG C.IV in Turkish markings after a landing accident. Many AEG C.IV aircraft were sent to the Middle East due to their robust steel-tube airframes, which absorbed the shock of this crash. Apparently no thought was given to the almost total lack of welding equipment in that remote, primitive theater because C.IV serviceability was greatly hampered by its absence. Even minor structural damage was very difficult to repair due to the scarcity of welding equipment.

Above: AEG C.IV 6648/16 of the second production batch with tactical number '5' served with *Flieger Abteilung(A)* 224, a Württemberg unit.

Left: AEG C.IV 6590/16 with checkerboard marking gets ready for take-off.

Right: An AEG C.IV crew ready for their next mission photographed before take-off. The large, unprotected radiator was vulnerable to ground fire and created significant drag. If it leaked due to damage the hot water spraying over the crew must have been very uncomfortable.

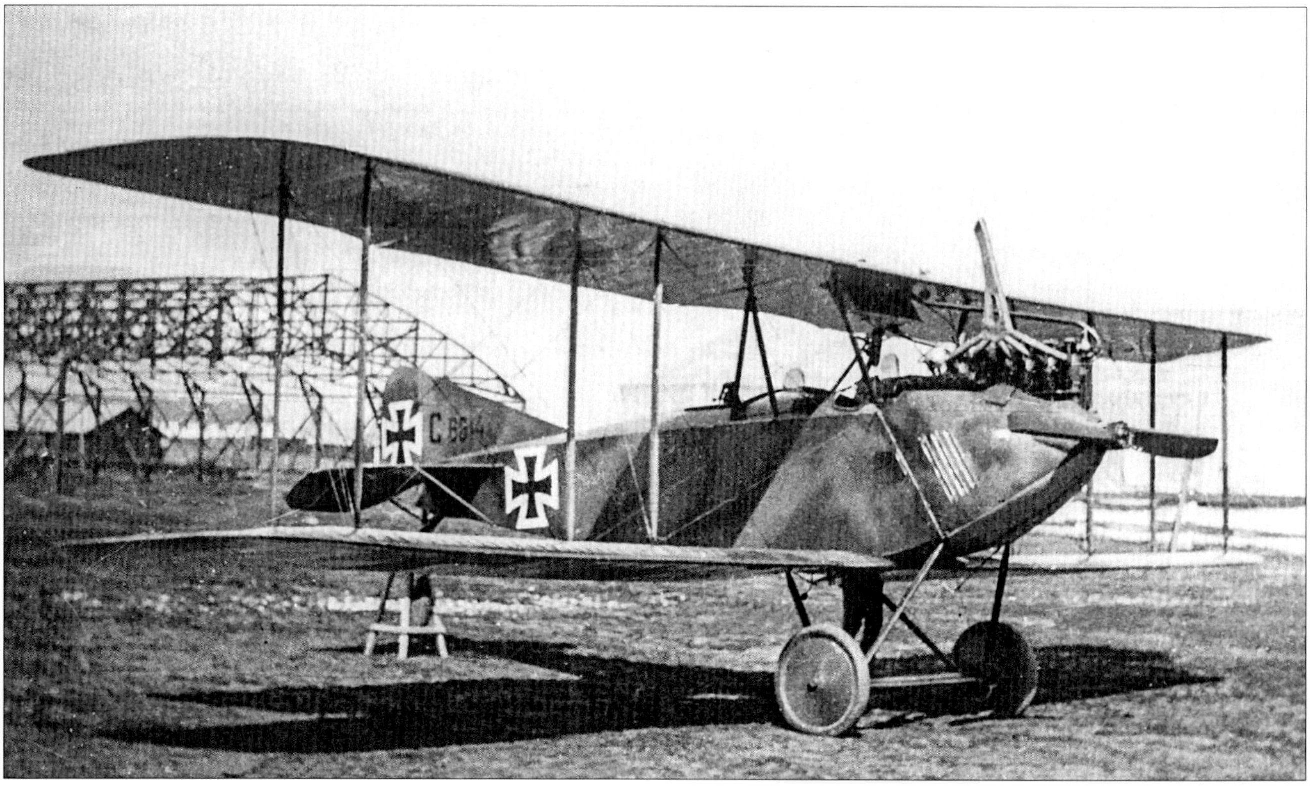

Above: AEG C.IV 6614/17 of the second production batch poses for its portrait.

Above: AEG C.IV with engine cowling panels removed shows details of the engine installation.

Right: *Vzfw*. Raetsch and *Offz*. Schulz in front of their C.IV.

Above: AEG C.IV with aircrew, one of whom is distinctly taller than the others, serving with *FA*304b in Palestine.

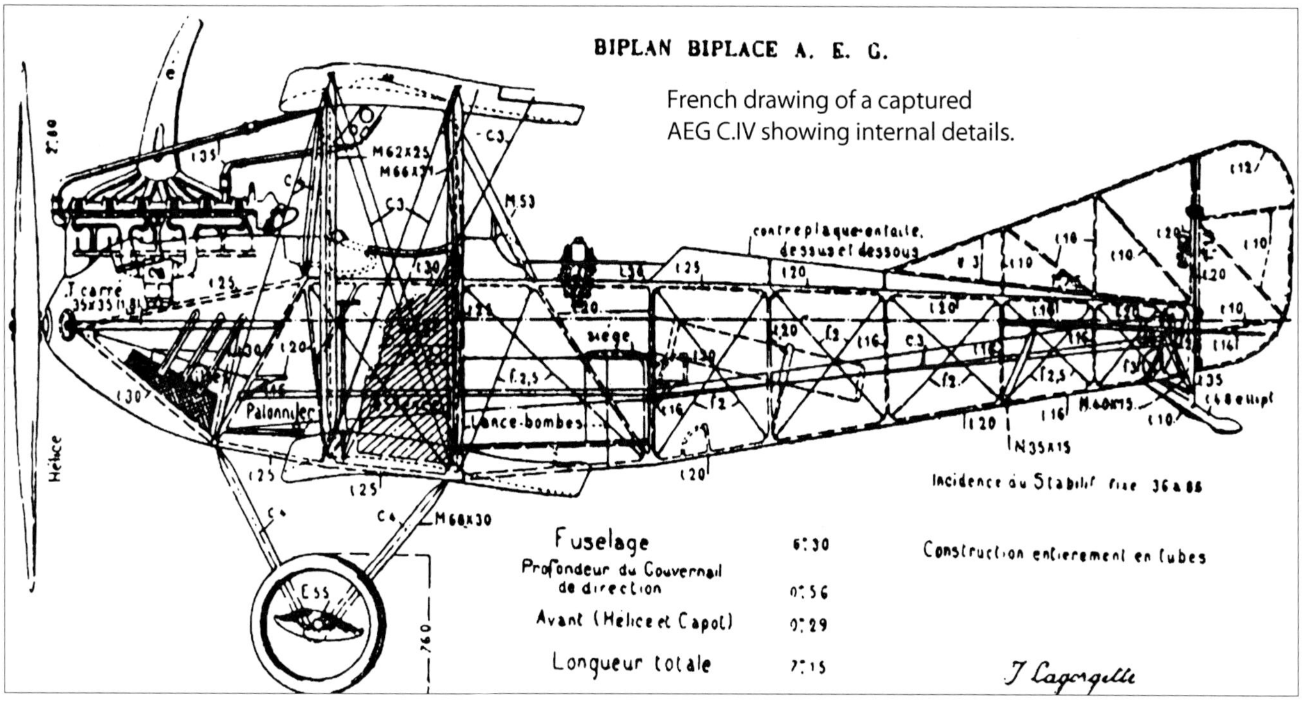

French drawing of a captured AEG C.IV showing internal details.

Above: AEG C.IV with the typical AEG two-color sprayed camouflage.

Above & Left: Landing accidents were common during WWI due to the rough fields and sensitivity of the light aircraft to gusts of wind while landing. These two AEG C.IV aircraft demonstrate some of the results. Above is C.IV 6602/16 of the second production batch.

Above: Sanke card 1048 featured the AEG C.IV(Fok) as shown by the Fokker-style streaked camouflage.

Below: Fokker-built AEG C.IV(Fok) 277/17 was assigned to the *Flieger Beobachter Schule Cöln* on Sep. 28, 1917 and landed in Heithuizen, the Netherlands, on Dec. 18, 1917. After repair and addition of the Dutch orange circle insignia it was assigned Dutch military serial AEG 403 and served in the Netherlands. The streaked finish of the Fokker-built C.IVs was very distinctive.

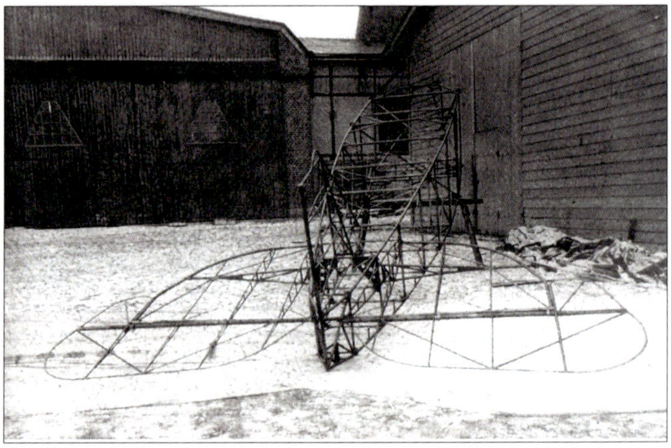

This Page & Facing Page: Structural details of the AEG C.IV(Fok) at the Fokker factory. Fokker and AEG were the only two German aircraft manufacturers that used welded steel tube structures. When Fokker's designs fell behind and were not being produced for the front, Fokker was given an order for license-built AEG C.IV aircraft to sustain the factory.

This Page: Additional structural details of the AEG C.IV(Fok) at the Fokker factory. Fokker-built AEG C.IVa aircraft were intended for training use; when Fokker received substantial orders for the Fokker D.VII fighter he requested the remaining AEG C.IVa(Fok) aircraft on the order be cancelled to free production capacity for the D.VII and this was granted.

Above: AEG C.IV(Fok) aircraft under construction in the Fokker factory.

Below: Wing rib details of the AEG C.IV(Fok).

Bottom: Undercarriage details of the AEG C.IV(Fok).

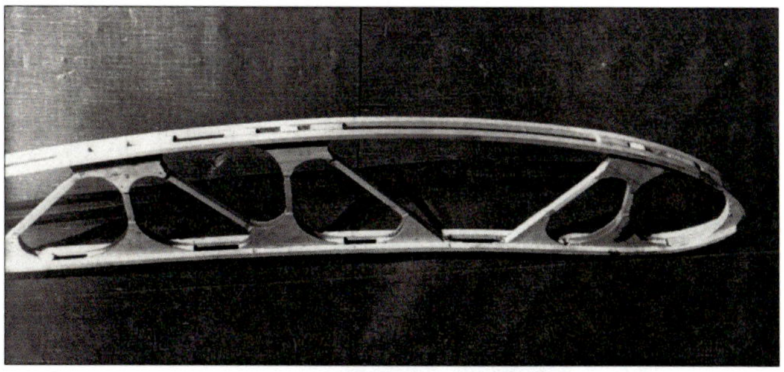

Below: Aileron covering details of the AEG C.IV(Fok).

Bottom: Structural details of the AEG C.IV(Fok).

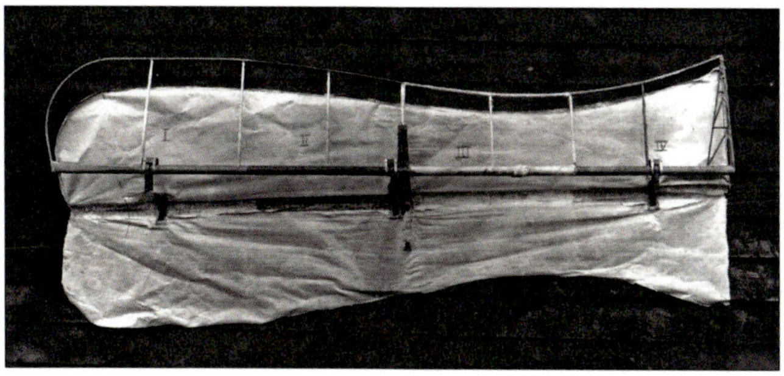

Above: *Lts*. Milenz and Spiegel with an AEG C.IV of *Flieger Abteilung* 303 in Palestine.

Below: AEG C.IV in Turkish markings at right and an Albatros C.III in Turkish markings at left.

AEG C.V

When a new engine became available it was common for *Idflieg* to order new prototypes from several manufacturers; this gave more experience with the new engine and enabled *Idflieg* to choose the best aircraft for production. When the new 220 hp Mercedes D.IV straight-eight became available *Idflieg* ordered new C-types powered by this engine from AEG, Albatros, and LVG. The resulting AEG C.V strongly resembled the AEG C.IV that was designed in parallel. Interestingly, the C.V was completed in February 1916, one month before the C.IV. Having a heavier engine, the C.V was slightly larger than the C.IV and was slightly faster. The competing Albatros C.V and LVG C.IV apparently had better performance or flying qualities than the AEG C.V because they were both ordered into production while the AEG C.V was not. Production of the Mercedes D.IV was limited and the engines allocated to AEG were used in the AEG G.III twin-engine bomber.

Above & Below: Despite its additional 60 hp, the AEG C.V was only slightly faster than the C.IV designed in parallel.

AEG C.VI

Apparently the AEG C.VI was an un-built project because there is no information about it in existing German records and no photographs exist of an AEG C.VI.

AEG C.VII

In August 1916 *Idflieg* issued a requirement for a two-seat escort aircraft. This was to be a lightened C-type, later categorized as a CL-type, powered by a 160–180 hp engine. AEG was one of the companies that responded with a proposal, and in October *Idflieg* gave AEG a contract for three prototype C.VII aircraft. The C.VII was powered by a 160 hp Mercedes D.III and the first aircraft was completed in December 1916. It was about 20% smaller than the C.IV powered by the same engine. Initial flight testing was apparently done in January 1917 at Nieder-Neuendorf, and the slightly modified C.VII prototype underwent testing in March. The initial *Idflieg* report (dated April 30, 1917) said the C.VII offered little that was new. The observer's field of fire was considered poor for an escort aircraft and the performance, while superior to the C.IV, was not as good as expected although flying qualities were acceptable. The C.VII was recommended for service on the less-demanding Eastern Front providing the load tests were passed. That recommendation was not compelling and no production was undertaken.

The third C.VII prototype was built with a highly-swept-back wing with minimum gap, intended to improve the gunner's field of fire. However, the competing Halberstadt and Hannover CL-types were superior and were placed in production while the AEG C.VII was not.

Above & Left: Either the first or second AEG C.VII prototype is shown here. Designed as an escort fighter, a category that later became the CL-class, the C.VII was acceptable but not superior, and the excellent Halberstadt CL.II and Hannover CL.II were produced instead.

Above & Below: The third AEG C.VII prototype featured a swept-back upper wing with minimum gap to improve the gunner's field of fire. However, the competing Halberstadt and Hannover CL-types were still superior aircraft.

AEG C.VIII & C.VIII Triplane

The AEG C.VIII was an attempt by AEG to improve the C.VII for the light two-seater escort role. Three prototypes were ordered, all powered by the 160 hp Mercedes D.III. Apparently two were completed as biplanes, the first in July 1917, and the third as a triplane. It should be remembered that the summer of 1917 was the height of *Idflieg's* triplane craze, which no doubt influenced AEG to complete one C.VIII as a triplane. AEG apparently hoped to achieve high speed by reducing drag to a minimum; the AEG history claims 190 km/h but engineering specifications show a top speed of 170 km/h. If the higher speed were actually achieved the C.VIII might have been ordered into production, but it was not and a reasonable conclusion is the 190 km/h top speed was not reached.

The C.VIII triplane used the engine, fuselage, and tail of the biplane C.VIII. Completed in October 1917 and flight tested in November, the C.VIII triplane climbed to 5000 meters in 34 minutes, a climb equivalent to the Albatros D.V single-seat fighter. With 80 kg less load it climbed to 5000 meters in 24 minutes. However, it was slower than the biplane C.VIII and flight-testing revealed the C.VIII triplane to have unsatisfactory flying qualities. AEG had proposed a smaller, lighter triplane in November but the disappointing flight-test results of the C.VIII triplane eliminated that idea as well.

Above & Below: One of the first two AEG C.VIII prototypes; the C.VIII was a cleaner, more aerodynamic design than the C.VII.

Above: The third AEG C.VIII prototype was built as a triplane. The climb rate was good but like most triplanes it was slower than the biplane from which it was derived, and it did not handle as well. (The Peter M. Bowers Collection/The Museum of Flight)

AEG Fighters

AEG only built two unarmored fighters, the D.I biplane and its Dr.I triplane. Both were fast aircraft but unexplained crashes unfortunately killed the very promising D.I program that might have given Germany technical superiority, and the speedy D.I over-shadowed its Dr.I triplane derivative.

AEG D.I

The AEG D.I fighter prototype, powered by a 160 hp Mercedes D.III engine and armed with two synchronized machine guns, was completed in May 1917. Like all AEG aircraft, the airframe was built totally from steel tubing. The wings had a single, steel-tube spar with wooden ribs, later changed to steel ribs, a feature that imparted great strength and minimized the number of drag-producing interplane bracing wires needed. With a top speed of 225 km/h (137 mph) and a climb equal to the Albatros D.V then just entering service, the AEG D.I prototype was the fastest fighter of its day. In June 1917, *Idflieg*, recognizing the fighter's potential, ordered three prototypes numbered D.4400–4402/17 for flight and static-load tests.

Between 28 June and 3 July 1917, as part of the *Typenprüfung* (type-test) at Adlershof, load tests demonstrated that the D.I wings were strong enough but the fuselage failed. After the reinforced fuselage passed its tests on August 4, 1917, the D.I was approved for flight evaluation by military pilots. At this time the fuselage of one prototype was lengthened by 40 cm to improve longitudinal stability and improve landing characteristics. On 21 August 1917, the "highly-skilled" *Idflieg* test pilot *Leutnant* Julius Hendrichs was killed when an AEG D.I went into an "ever steeper dive" and crashed out of control from 400 meters (1,310 ft). The cause of Hendrichs' crash was not determined. Using another D.I, the flight evaluation and engineering critique was completed August 25, 1917. Provided some minor installation issues were corrected, the AEG

AEG D.I Specifications

Engine:	160 hp Mercedes D.III	
Wing:	Span Upper	8.50 m
	Span Lower	8.14 m
	Chord Upper	1.38 m
	Chord Lower	1.07 m
	Gap	1.53 m
	Area	17.1 m^2
General:	Length	6.15 m
	Height	2.54 m
	Empty Weight	685 kg
	Loaded Weight	945 kg
Maximum Speed:		225 kmh
Climb:	1000m	2.5 min

Above: The AEG D.I prototype, 4400/17, photographed at the Nieder-Neuendorf airfield in May 1917. Because the narrow airfoil section did not allow the installation of an airfoil radiator, a nose radiator was installed. The long exhaust stack reduced exhaust noise.

Below: AEG D.I 4401/17 has a longer fuselage than the prototype for better stability. Ear radiators are fitted and the exhaust stack is much shorter than the prototype. The German fighter pilots needed an aircraft with the speed of the AEG D.I, which was faster than any operational German fighter of the war, so the decision to cancel it despite two fatal accidents must have been difficult. In retrospect it seems that continued development was warranted due to the aircraft's potential unless there was a fundamental and irreparable flaw in the design, which does not appear to be the case. The cause of the accidents was never determined so the potential for making the aircraft safer through further development is unknown. However, the small number of production aircraft were flown for some time, including as comparison aircraft in the First Fighter Competition, without further fatalities, bringing its cancellation into question.

Above: AEG D.I 4400/17 now has side radiators. It is shown here on Marville airfield shortly before *Lt.* Höhndorf's fatal flight in it on Sept. 5, 1917 that essentially ended D.I development despite its high speed.

D.I was approved for operational service. In July 1917 *Idflieg* had ordered 20 pre-production fighters numbered D.5000 to 5019/17 contingent on the type-test outcome.

Leutnant Walter Höhndorf, *Staffelführer* of *Jagdstaffel* 14 and a *Pour le Mérite* ace with 12 victories, had been involved in the design of the AEG D.I. An engineer and licensed pilot, Höhndorf had previously worked as a designer and test pilot at the Union Flugzeugwerke. The chief engineer at Union was Georg König, now at AEG. On September 5, 1917, Höhndorf was killed when the modified AEG D.I 4400/17 spun out of control and crashed. With two unexplained fatalities blemishing its record, *Idflieg* canceled the AEG D.I program after a few of the pre-production fighters were completed.

Above: AEG D.I 4400/17 now has side radiators and a shorter exhaust stack. It is shown here on Marville airfield shortly before *Lt.* Höhndorf's fatal flight in it on Sept. 5, 1917 that essentially ended D.I development despite its high speed.

AEG Dr.I

The AEG Dr.I was built at the height of Germany's 'triplane craze' as a triplane version of the D.I biplane. Completed in October 1917, like the D.I it was powered by a 160 hp Mercedes D.III and was armed with two synchronized machine guns. Ailerons on the top and bottom wings were connected by actuating struts. The Dr.I had greater wing area than the D.I and demonstrated a slightly better climb rate. However, it was also heavier and had more drag, so, according to *Idflieg*, "the speed was drastically lower, precluding any consideration of front-line service." However, the AEG Dr.I was still faster than any other German fighter design at that time – except the AEG D.I biplane – and was the fastest German triplane fighter.

AEG Dr.I Specifications

Engine:	160 hp Mercedes D.III	
Wing:	Span Upper	9.40 m
	Span Middle	9.10 m
	Span Lower	8.80 m
	Chord Upper	1.37 m
	Chord Middle	0.75 m
	Chord Lower	0.75 m
	Gap Upper	1.68 m
	Gap Lower	0.75 m
	Area	24.0 m^2
General:	Length	6.15 m
	Empty Weight	710 kg
	Loaded Weight	970 kg
Maximum Speed:		200 kmh
Climb:	1000m	2.5 min

Above: The AEG Dr.I was built to compare a triplane development of the AEG D.I biplane with the original biplane design. The AEG Dr.I appeared during the height of the German 'triplane craze' when it was common to build triplane versions of biplane designs. The climb rate was slightly improved due to the larger wing area but, as was nearly universal with these triplane versions, it had more drag and was significantly slower than the D.I biplane. Its lower speed compared to the AEG D.I eliminated any chance of production despite the fact the AEG Dr.I was faster than any German fighter then in service. (The Peter M. Bowers Collection/The Museum of Flight)

Above: Front view of the AEG Dr.I prototype highlights its clean design for a triplane. The strength of its wings due to steel spars meant that only minimum bracing wires were needed. Interestingly, the bracing wires went through the middle wing but did not connect to it. (The Peter M. Bowers Collection/The Museum of Flight)

Above: Here the AEG Dr.I wing cellule has been rebuilt to evaluate the wing design for the AEG DJ.I armored triplane fighter.

Below: Only one prototype of the AEG Dr.I was built. Ailerons on the top and bottom wings were connected by a strut.

AEG Armored Aircraft

AEG J.I

Above: AEG J.I J.153/17 in factory sprayed camouflage finish at Adlershof for type testing clearly shows its derivation from the earlier AEG C.IV. Changes to the C.IV to create the J.I included replacing the 160 hp Mercedes D.III with the more powerful 200 hp Benz Bz.IV, addition of armor plate around the cockpit and engine, elimination of the fixed, forward-firing gun for the pilot, and addition of ailerons to the lower wings for improved maneuverability.

The successful AEG C.IV, with its steel-tube airframe, was well-suited as a starting point from which to develop an armored J-type. To develop the new type quickly, flat armored sheets, three on each side and three on the bottom, were bolted directly onto the metal airframe of the C.IV. The armor weighed a total of 400 kg. Integrating the armor plates as structural elements would have saved some airframe weight but cost a lot of time to develop and manufacture, and the AEG J.I was needed at the front as soon as practical. To improve maneuverability with the greater weight, ailerons were added to the lower wings. Finally, to haul the extra weight into the air, a 200 hp Benz Bz.IV engine replaced the 160 hp Mercedes D.III engine used in the C.IV. At first the wing area was unchanged from the C.IV, and the additional weight raised the stall speed and lengthened take-off runs.

Assembly of the AEG J.I prototype began in January 1917 and was completed in April. The first five production J.I aircraft were delivered in May. On 30 May 1917, AEG test pilot Theodor Schauenburg was killed when he attempted to loop a J.I and the wings came off. What possessed him to perform this terminally-foolish maneuver is unknown because aerobatics understandably were not a requirement for J-types.

The J.I airframe was identical to the AEG C.IV that had already passed its type-test, so *Idflieg* delayed the static-load test (on J.150/17) until August 1917 to expedite deliveries to the front. This was long after the J.I had entered combat, which was unusual for German aircraft. Starting in September 1917 the first running changes were introduced into J.I production to decrease vulnerability to ground fire. These changes were installation of double lift-bracing wires and double control cables. The AEG J.Ia engineering prototype was used to evaluate additional changes that were introduced as production progressed (see table on page 11).

Because the AEG J-types were both relatively easy to built as well as resistant to ground fire, the AEG J.I and J.II became the most numerous J-types at the front. Like other J-types, the AEGs were assigned in ones or twos to divisional *Flieger Abteilungen* to fly contact patrol sorties, or *'Iffeln.'* In the field, the AEG J.I was complimented for its high speed, estimated to be 150–200 km/h in level flight and as high as 250 km/h in a shallow dive. Unfortunately, this was optimistic; actual top speed in level flight

AEG Armored Aircraft Specifications					
	AEG J.I	**AEG J.II**	**AEG DJ.I Triplane**	**AEG DJ.I Biplane**	**AEG G.IVk**
Engine	200 hp Benz Bz.IV	200 hp Benz Bz.IV	195 hp Benz Bz.IIIb V-8	195 hp Benz Bz.IIIb V-8	Two 260 hp Mercedes D.IVa
Span, Upper	13.00m	13.00m	11.20 m	10.00 m	18.00 m
Span, Lower	12.48m	12.48m	10.00 m	9.90 m	17.50 m
Chord, Upper	1.65m/1.85m	1.85m	1.475 m	1.80 m	2.40 m
Chord, Lower	1.65m/1.85m	1.85m	0.825 m	1.60 m	2.40 m
Gap	1.95m	1.95m	1.00 m	2.00 m	2.60 m
Wing Area	39.0 m^2/43.6 m^2	43.6 m^2	31.3 m^2	30.5 m^2	65 m^2
Length	7.20m	7.86m	—	6.90 m	9.00 m
Height	3.3m	3.3m	—	—	—
Track	2.19m	2.15m	1.90 m	1.90 m	4.80 m
Empty Weight	1,456 kg	1,480 kg	1,182 kg	1,185 kg	2,700 kg
Loaded Weight	1,876 kg	1,900 kg	1,412 kg	1,375 kg	3,150 kg
Maximum Speed	140 km/h	140 km/h	166 kmh	180 kmh	150 kmh
Climb to 1,000m	6.1 minutes	6 minutes	5.8 minutes	4 minutes	9 minutes
Climb to 2,000m	15.9 minutes	14 minutes	—	—	—
Climb to 3,000m	31.6 minutes	30 minutes	—	—	—
Climb to 4,000m	—	—	48 minutes	—	—
Armament	1 flexible machine gun, some had 2 fixed machine guns	1 flexible machine gun, 2 fixed machine guns	2 fixed machine guns, 4 small bombs	2 fixed machine guns, 4 small bombs	2 flexible machine guns, 2 flexible 20mm Becker cannon, small bombs

Notes: 1.AEG J.I specs are shown as early/late; later production AEG J.I aircraft had 1.85m chord and wing area of 43.6 m^2. AEG J.I dihedral = 2°
2. At least 20 J.II aircraft fitted with a flexible 20mm Becker cannon in place of the 2 fixed machine guns.
3. The middle wing of the AEG DJ.I Triplane had aspan of 10.60 m and a chord of 0.825 m.

was a mere 140 km/h. Due to its greater wing area and lighter weight, the Albatros J.I had a shorter take-off run, enabling it to operate from muddy airfields. In contrast, the heavy AEG J.I required a longer take-off run from drier surfaces. Though reasonably fast for an armored two-seater, it was not a fast airplane. Furthermore, the AEG J.I was cumbersome and difficult to land because the short fuselage and heavy nose made it easy to flip over on landing. On the other hand, crews felt more secure in the AEG J.I than the Albatros J.I because, unlike the Albatros, the engine was protected by armor. The armor's effectiveness was demonstrated by British firing tests on the armor plate of a captured J.I; results are given in the table on page 55.

Installation of a 20mm Becker cannon in an

AEG J-Type Orders & Production Notes

- About 135 AEG J.I aircraft were built; the rest were AEG J.II, making the AEG J.II the most numerous J-type.
- 508 AEG J-types were built through October 1918, and 609 were built through December 1918. About 373 and 474 respectively were J.IIs.
- The AEG J.III prototype was either J.375/18 or J.575/18.
- The AEG J.II series J.375–574/18 is conjectural. It is included because of the 609 AEG J-types built by the end of 1918; otherwise there are not enough serial numbers for these aircraft. Also, AEG J.II J.417/18 was used as a civil aircraft in 1919, proving that some aircraft in this serial number range were built.

AEG D.I 4400/17

AEG Dr.I

AEG C.IV C.6623/16

Right: Front view of AEG J.I J.153/17 showing the factory sprayed camouflage and the lead for the wireless antenna hanging under the fuselage. Ailerons on all four wings provide added maneuverability to the heavy aircraft, but no horn balances are fitted to this early-production J.I.

Below: Early production AEG J.I J.151/17 at *FA(A)* 250.

Below: AEG J.I J.220/17 *Margo* of *Flieger Abteilung (A)* 208 wears the unit's reverse swastika marking.

Above: AEG J.I of *Schlasta* 26b wears that unit's red flames marking on its nose.

Above: AEG J.I J.236/17 wears sprayed camouflage with late German insignia and has two downward-firing guns.

Above: *Flieger Abteilung (A)* 208 lineup at Voyenne in the Spring of 1918 shows a variety of two-seaters with that unit's reverse swastika marking on their fuselages, including an AEG J.I at center.

Below: A closeup of the above photo of *Flieger Abteilung (A)* 208 shows the AEG J.I at right, with a captured Breguet 14 at left. Was the AEG J.I at right *Margo* as shown in the photo on the previous page?

AEG J.I fitted with an enlarged gun ring was undertaken in January 1918, followed by firing trials in February–March. The firing trials demonstrated the cannon was difficult to operate in the air as mounted, making this type of installation unsatisfactory for two-seaters.

Former observer Hanns-Gerd Rabe flew the AEG J.I in combat. "In order to make close observation

AEG J-Type Modification Summary

AEG introduced running changes into production of their J-types; the key changes are summarized here.

Changes from AEG C.IV to Create AEG J.I

- Change engine from 160 hp Mercedes D.III to 200 hp Benz Bz.IV
- Ailerons added to lower wings for better maneuverability
- Armor added to protect engine and crew
- Pilot's fixed gun deleted

Running Changes Introduced During AEG J.I Production

- Double lift-bracing wires and double control cables installed to increase resistance to ground fire (tested on AEG J.Ia)
- Two fixed, downward-firing guns added for more offensive capability (also used in AEG J.II)
- Lengthened fuselage for improved stability, especially during landing
- Enlarged wing area for shorter takeoff run by increasing wing chord (tested on AEG J.Ia)
- Improved aileron control linkage for better maneuverability (tested on AEG J.Ia)

Changes Introduced to Create AEG J.II

- Front spar composed of two adjacent tubes to reduce susceptibility to ground fire
- Horn-balanced ailerons (upper wings) for improved controllability (various configurations used)

Changes Introduced During AEG J.II Production

- Wing spars, fuselage frames, and aileron control horns strengthened
- Horn balances added to elevators and rudder for improved maneuverability
- 20 J.II aircraft built with 20mm Becker in floor, Sept. 1918
- All J.II aircraft had two downward-firing guns except those with 20mm Becker in floor

Changes Introduced to Create AEG J.III

- 260 hp Mercedes D.IVa replaced 200 hp Benz Bz.IV for better performance
- Redesigned, lightened armor plate for better performance

Results of British Testing on AEG J.I Armor Plate Effectiveness

Ammunition	Angle to Normal, °	Safe Range, Yards	Unsafe Range, Yards
German A.P.	0	—	600
	15	500	400
	30	400	300
Mark VII.P Armor Piercing (angle to normal approximate)	0	700	600
	15	400	300
	30	300	200
German Spitze	0	150	100
	15	100	50
	30	50	50
Mark VII	0	50	—
	15	50	—
	30	50	—

British authorities performed an engineering analysis of an AEG J.I brought down on 16 May 1918. During the course of the analysis, the ability of various types of ammunition to penetrate the 5mm thick armor at various ranges and angles was tested, resulting in the data tabulated above. The armor was judged to be too light to protect against British armor piercing ammunition below 500 feet altitude.

Above: Typical early production AEG J.I with wireless antenna lead visible below the observer's cockpit.

possible and to take part in the land battles, armored airplanes were built with the engine, fuel tank, and pilot well-protected by armor plate. The observer was less protected because he had to move about freely to make his observations. Equally unprotected were the radiator above the engine and the control cables. At the end of July 1918 my unit, *Flieger-Abteilung (A)* 253, received an armored airplane of the AEG J.I type. We regarded it with suspicion for it was too heavy and seemed to be clumsy as the 200 hp engine was too weak. The pilots agonizingly circled the airfield with it, but it barely went over 500 meters. Risky maneuvers in aerial combat were out of the question... In our opinion it was a very primitive bird!" This was the opinion of many combat aircrewmen, who clearly did not agree with Idflieg's theories on required J-type performance.

Continuing Rabe's comments, "The take-off was laborious; the crate lifted off very sluggishly and hung in the air... I heard the machine gun bullets hitting the armor plating and also saw holes being torn in the fabric... but I wasn't really concerned... When I had exposed all my photographic plates, I threw my hand grenades from the lowest altitude into the English posts and dropped gas bombs before the entrances of the dug-outs, and then fired my machine guns. I could work both machine guns in the floor with a foot pedal and I also fired the rotating machine gun at the machine gun emplacements in the trenches that were mainly firing at me from the rear... Suddenly, hot water hit me in the face. The vulnerable radiator had been hit. We could not last for more than a few minutes before the engine over-heated and froze up. Peter Johannes (the pilot) turned the AEG in a flat curve... to take us back to our own lines. But in the banked position, hits on the control cables to the rudder and ailerons tore them apart and caused the machine to dive towards a thick mass of barbed wire... Of course, the 'Emil' (nickname given to German two-seater pilots; the observer was 'Franz') succeeded in once again laboriously bringing the AEG back on an even keel, but the landing gear got caught in the barbed wire and the machine rolled over, leaving the landing gear pointed skyward..."

According to a French intelligence bulletin dated November–December 1918, the fuselage of an AEG J.I 468/17 (*lang*) was found near Strasbourg. The J.I was prone to landing accidents (flipping over) because of its short fuselage and heavy nose armor, and lengthening the fuselage for better pitch stability and control was an appropriate modification. The number of AEG J.I aircraft manufactured with longer fuselages is not known.

Above: AEG J.I J.203/17 had the doubled lift wires introduced into production in September 1917 to improve survivability. Identification streamers, probably in red but black was also used, attached to both rear inner interplane struts, are hard to see because the wind is whipping them.

AEG J.I J.203/17

AEG J.I of *Schlasta* 26b

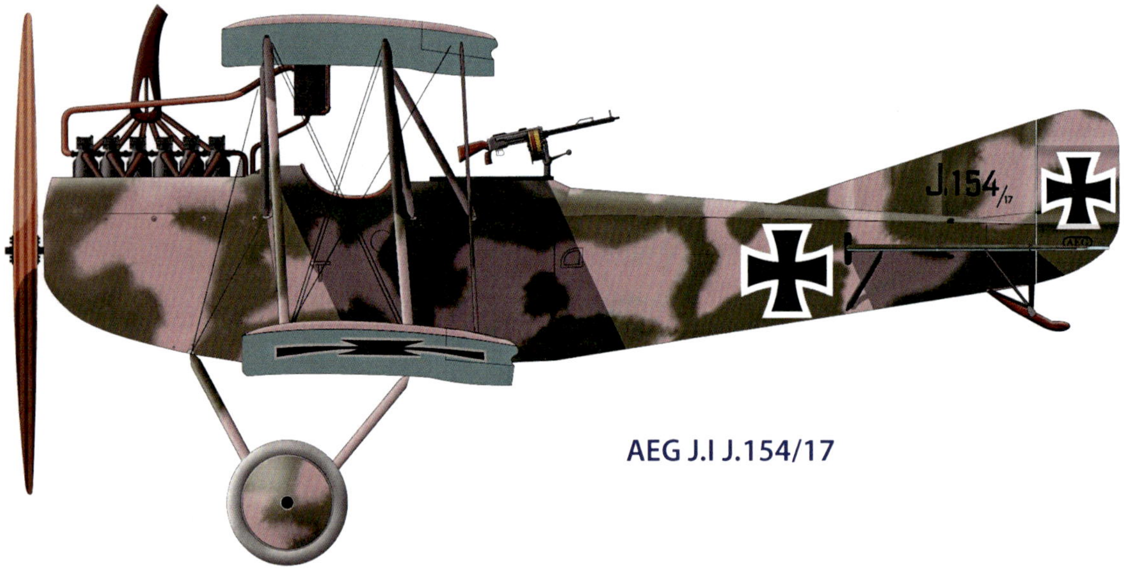

AEG J.I J.154/17

AEG J.I J.220/17 *Margo* of *Flieger-Abteilung (A)* 208

Above & Below: Two views of early-production AEG J.I J.154/17 wearing its sprayed camouflage pattern.

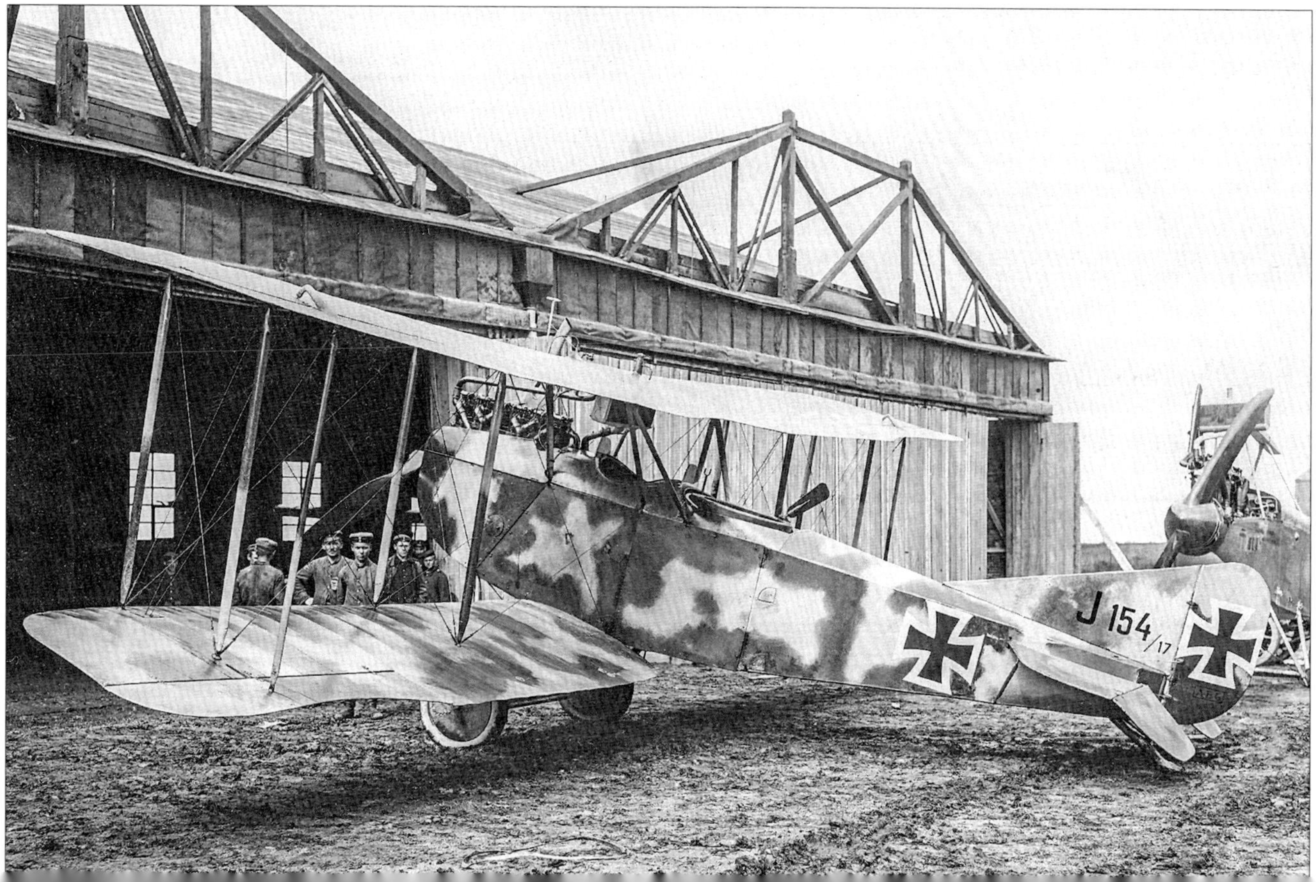

AEG J.I J.216/17 of *Flieger-Abteilung (A)* 221

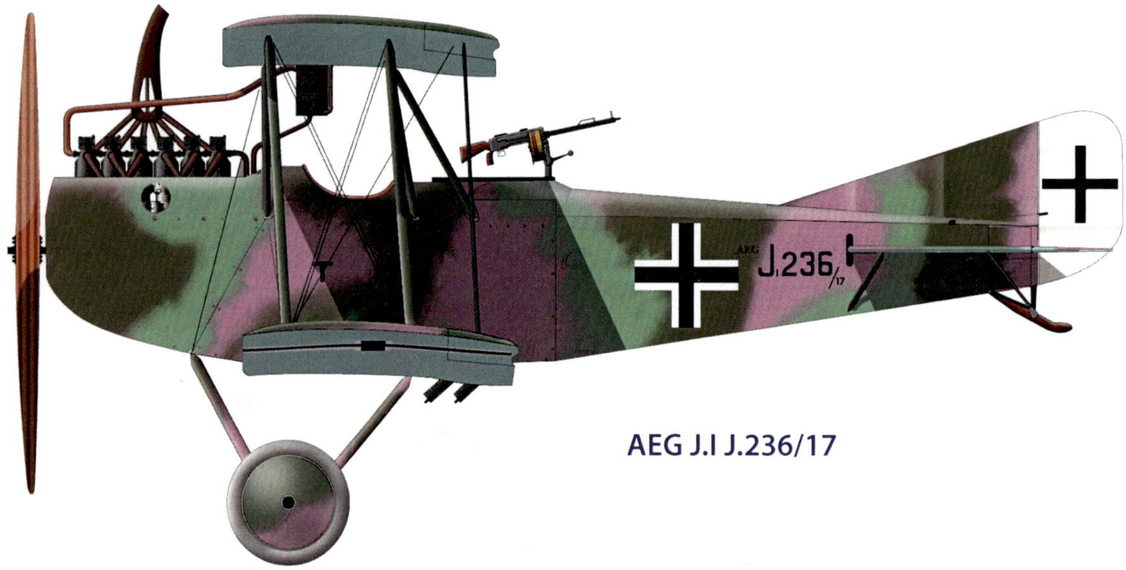

AEG J.I J.236/17

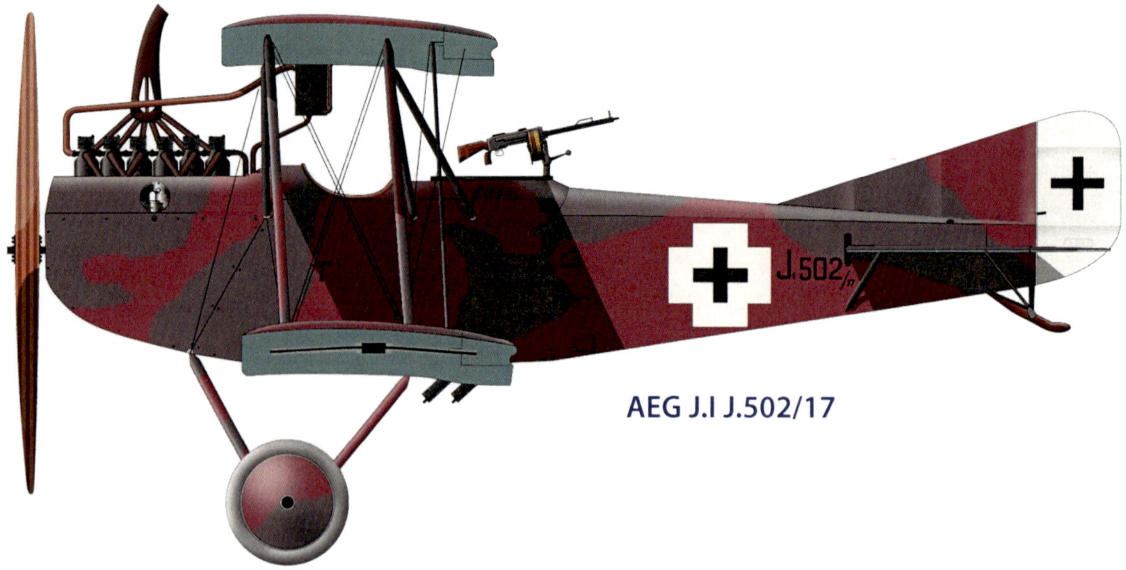

AEG J.I J.502/17

Above: Early-production AEG J.I J.158/17 in sprayed camouflage finish photographed at the AEG company airfield at Nieder-Neuendorf. The attachment to the gun ring is to counterbalance the drag of the machine gun, reducing the force the gunner needed to apply to turn the gun against the slipstream.

Above: An unidentifed AEG J.I.

Above & Below: Two views of early-production AEG J.I J.159/17 at *Flieger Abteilung (A)* 255 wearing its factory-applied sprayed camouflage pattern but no unit or individual markings.

Above: Later-production AEG J.I J.502/17 with downward-firing guns, straight-sided insignia, and later camouflage. This aircraft has the relocated aileron connecting struts as tested on the AEG J.Ia.

Above: Early-production AEG J.I with early sprayed camouflage pattern at *Flieger Abteilung (A)* 250. Red or black identification streamers are folded on the lower wing.

Above: The AEG J.Ia was a developmental aircraft to test engineering changes for the AEG J.I. Among the changes was enlarged wing area achieved by use of a greater wing chord, modification of the aileron control linkage for better maneuverability, and double lift wires and control cables to make the J.I less susceptible to fatal damage from ground fire. All these changes were introduced to the AEG J.I during its production run and were continued in the AEG J.II.

AEG J.Ia

AEG engineers began work in August 1917 to improve the J.I design. The wing loading was reduced by enlarging the wing area by 12 percent to reduce the stall speed and shorten the take-off run. The wing span remained constant; the wing area was enlarged by increasing the wing chord from 1.65 meters to 1.85 meters. To reduce the pilot's control forces, which improved maneuverability, the strut connecting the upper and lower ailerons was moved and attached to the aileron linkage. Installing double lift wires and control cables made the aircraft less vulnerable to ground fire. Some of these modifications, which were tested in August–October 1917 on the AEG J.Ia, were gradually introduced on the J.I production line, and later on all of the J.II machines. AEG J.Ia was not a military designation and appears to have been an internal AEG designation for a single development aircraft.

AEG J.II

As flight testing of the AEG J.Ia continued, the design of the AEG J.II, which incorporated many J.Ia features, was nearing completion. To further reduce vulnerability to ground fire the J.II's front spar was composed of two parallel steel tubes. By October 1917 the J.II prototype was undergoing flight evaluation. In January 1918 the flight-test results were still unsatisfactory, especially the poor rate of climb. A new control wheel for the pilot gave control forces that were much too heavy, hindering maneuverability.

The evolution of J.II control surfaces during its production run is evidence that flying qualities needed improvement, but production could not be

Above: AEG J.II J.186/18 is an early production J.II without balanced control surfaces, making it look like a late-production J.I. The aircraft did have the larger wing chord and pair of fixed, downard-firing guns standard in the J.II.

Above: AEG J.II J.232/18 is another early production J.II with unbalanced control surfaces, making it look like a late-production J.I. Like the forward fuselage, the underwing radiator was armored.

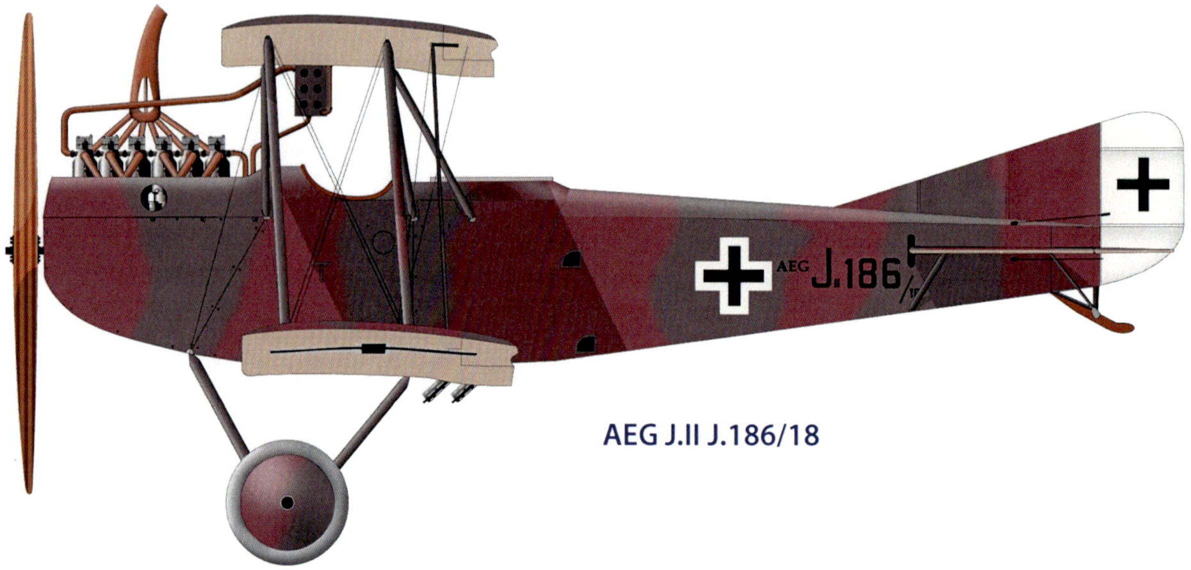

AEG J.II J.186/18

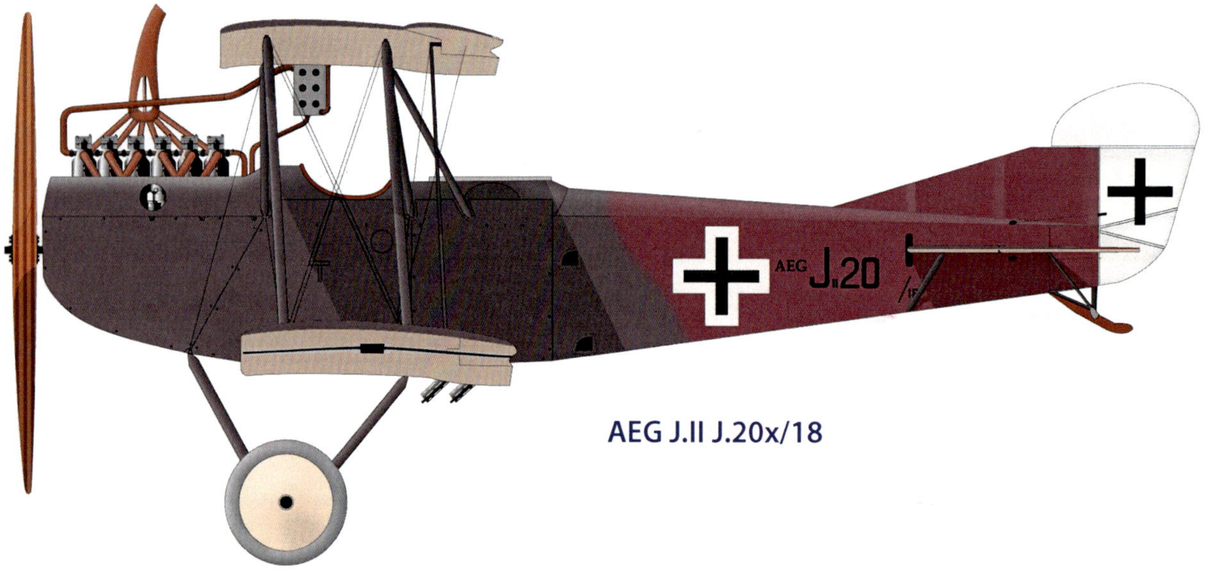

AEG J.II J.20x/18

AEG J.II J.306/18

Above: Factory photo of AEG J.II J.20x/18 with ailerons fitted with inset aerodynamic balances on upper and lower wings. Balanced tail surfaces are also fitted. The pair of fixed, downward-firing guns are clearly visible.

Above & Below: An AEG J.II, possibly J.294/18, at Adlershof in August 1918. This represents the final production configuration with all control surfaces fitted with horn aerodynamic balances except the ailerons on the lower wings. The pair of fixed, downward-firing guns are visible, and all fabric surfaces are covered with five-color printed camouflage fabric.

Above: AEG J.II J.306/18 is a in final production configuration and covered with four-color hexagonal camouflage fabric overall except for the fuselage armor, which is painted a solid dark color.

interrupted because aircraft were needed at the front. The first J.II production machines were delivered with unbalanced control surfaces. Flight testing of aerodynamically-balanced controls was especially time-consuming because it was totally empirical. Furthermore, flight testing was delayed by winter weather that limited test-flight opportunities. A few early production aircraft featured balanced ailerons within the upper wing planform, but the standard configuration had projecting horn-balances for all control surfaces except the lower wing ailerons.

The AEG J.II was in quantity production by March 1918, and by April it was in combat at the front. Because of its similarity to the J.I, the static load test was apparently waived to expedite deliveries. According to combat reports, the AEG J-types performed very well during the Allied summer offensives. However, the long take-off run was still criticized. In fact, *Idflieg* considered installing the 260 hp Mercedes D.IVa engine to improve take-off performance, something that should have been done in the beginning. However, no changes were introduced, probably due to limited supplies of this popular engine. In fact, failure to employ the powerful 260 hp Mercedes D.IVa is inexplicable except as a consequence of inadequate supplies.

Several J.II crashes compelled *Idflieg* to perform a full static load test, which was done using J.II 294/18 during 8–20 August 1918. Testing revealed that in a shallow dive (Case C) the wing spar joints failed at 64% of the required load. Some fuselage frames and the aileron controls also needed strengthening. The problems were traced to a combination of poor materials, an increasing problem throughout the aviation industry at this stage of the war, and deficient manufacturing processes.

Apparently all J.II aircraft were delivered with the two-gun battery mounted in the observer's cockpit except for those with a 20mm Becker cannon. In June 1918, a single machine-gun was experimentally fitted in the nose so the pilot flying at 100 meters could fire at ground targets about 460 meters ahead. However, this was not placed in production. Also in June 1918, a flexible 20mm Becker cannon installed in the floor of the observer's cockpit was tested. Slotted armor grating protected the gun opening and aiming window. In September 1918, AEG delivered 20 J.II aircraft armed with the 20mm Becker cannon, primarily for destroying tanks.

Because of production with no recorded break between types, the exact number of AEG J.I and J.II aircraft built is unknown, but about 135 J.I and the remainder J.II is a reasonable estimate. Up to October 1918 some 508 J-types were delivered by AEG, more than any other manufacturer, which agrees with the Inter-Allied report that a total of 609 AEG J-types were built, a total that includes the machines completed in November and December 1918 and later adapted for civil use.

Facing Page: Armor plate has been removed from AEG J.II J.241/18 to show its pair of fixed Spandau machine guns mounted in the rear cockpit. The guns were offset to the left to leave space for the gunner's seat on the right.

J II 241

Above: The running engine of this late-production AEG J.II does not seem to disturb the crewmen in the photo.

Above: Late-production AEG J.II J.270/18 has come to grief, apparently the result of a bad landing at its aerodrome. The pair of downward-firing machine guns is visible protruding through the armor plate.

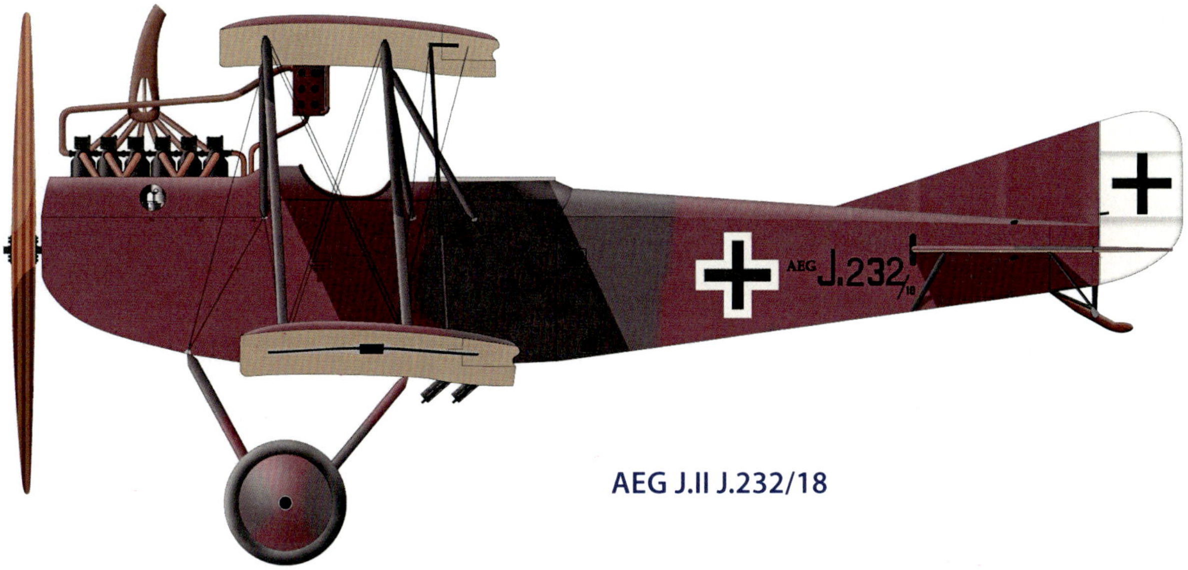

AEG J.II J.232/18

AEG J.II J.336/18

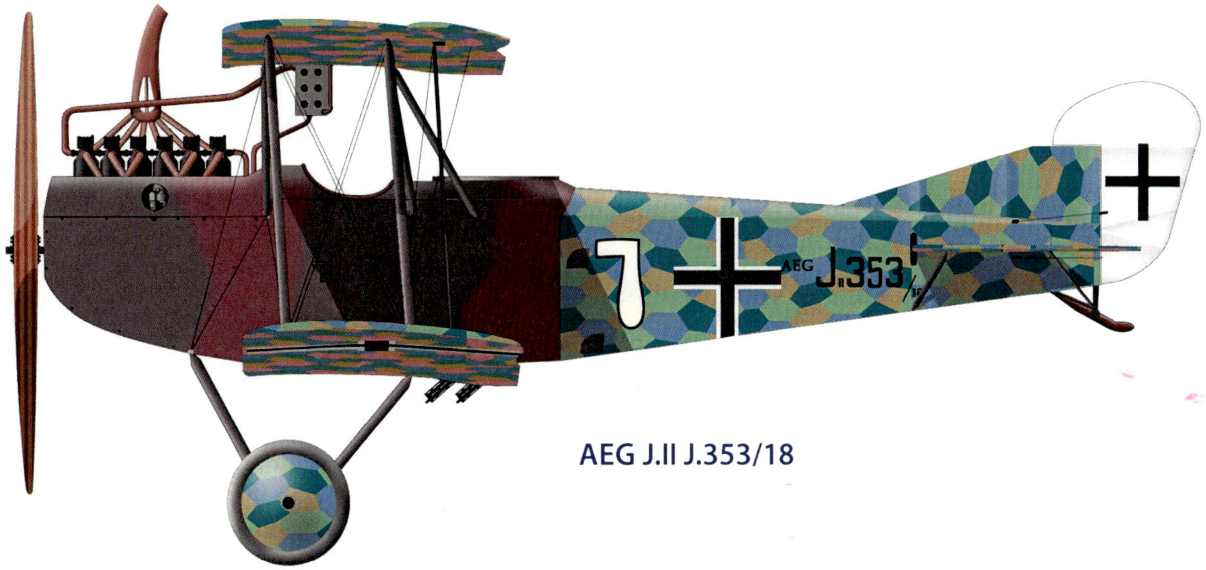

AEG J.II J.353/18

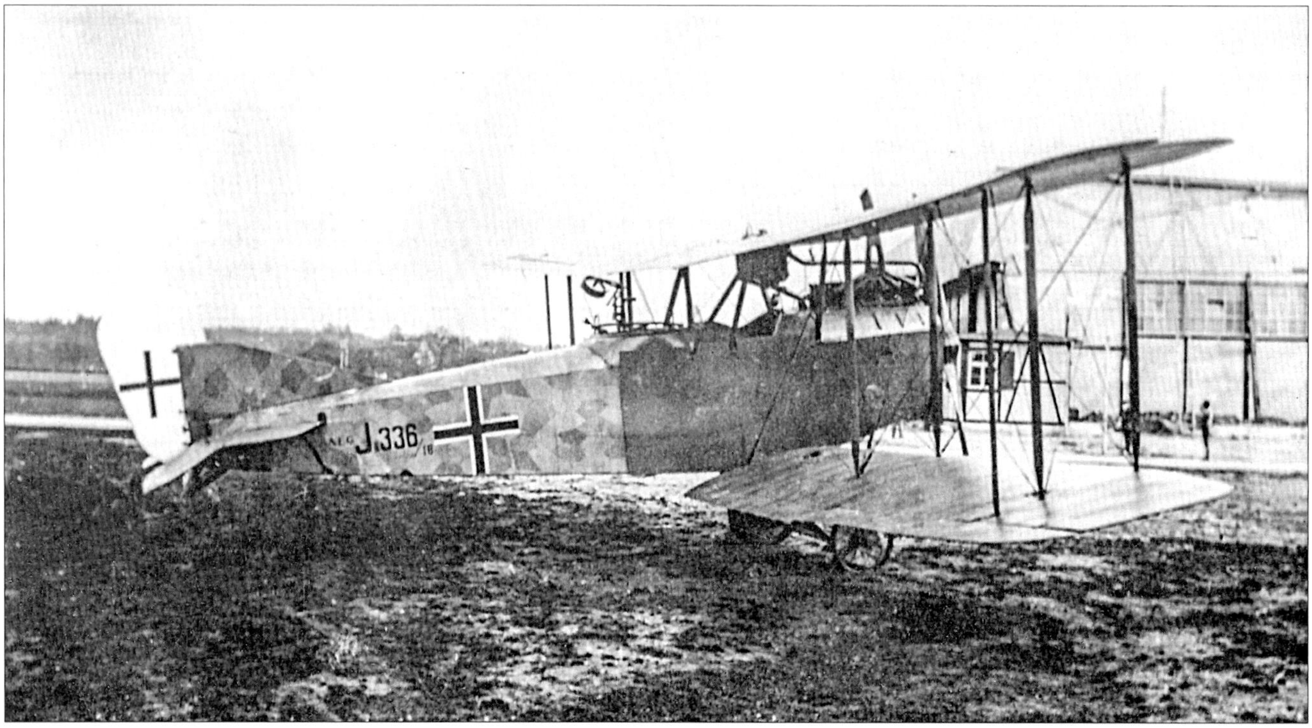

Above & Below: Late-production AEG J.II J.336/18 is shown in standard factory camouflage and markings.

Above: Another AEG J.I from *Fl Abt. (A)* 208. It also had the white swastika but was named "*Jumbo*".

Below Left: Mounting for twin downward-firing guns in AEG J.I and J.II.

Below: Observer's triggers to fire the two downward-firing guns in the AEG J.I and J.II. Both drawings from *The Engineer* of Sept. 6, 1918.

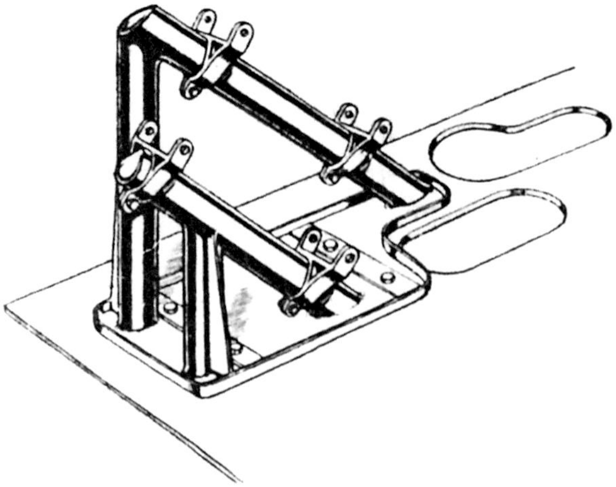

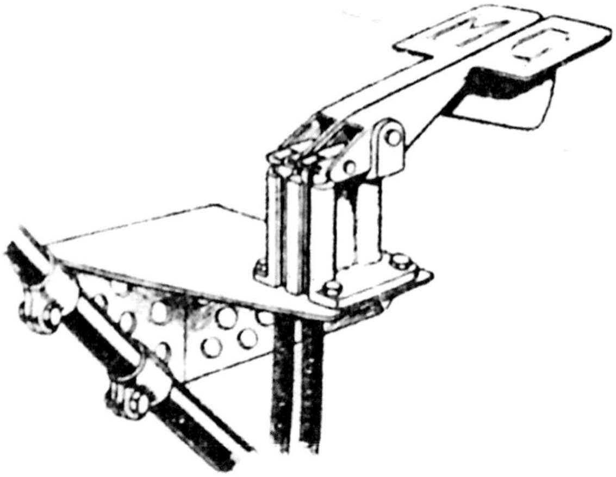

Above: Late-production AEG J.II J.336/18 is shown in standard factory camouflage and markings. Unusually, the downward-firing guns are not present.

Right: The wreck of AEG J.II J.249/18 brought down in US lines on 27 October 1918. The location was said to be ½ mile southeast of Cunel. Details of the armor are visible, including where it was bolted to the airframe. There were three plates on each side and another three under the fuselage. This photo shows the downward-firing machine guns clearly. The hole in the armor just aft of the starboard undercarriage strut enabled the gunner to view the ground beneath the aircraft and fire when the guns were pointing at a target. Arrangements for the downward-firing guns in the AEG and Albatros J-types were similar.

Facing Page, Top: The fuselage of AEG J.II J.353/18 is shown in the huge storage facility in Romorantin, France, where many German aircraft were stored post-war. It was tactical number '7' in its unit. AEG J.II 353/18 with the white "7" on the fuselage is also visible in photos of the Zeppelin hangar at Trier (Treves in English and French) too, indicating that it was turned over to the Allies at Trier, Germany, then transported to Romorantin, France.

This Page: This is a series of photos of the wreck of AEG J.II J.249/18 brought down in US lines on 27 October 1918. The location was said to be ½ mile southeast of Cunel; the captions on the photos indicate it was downed by a fighter. Details of the armor are visible and the two bottom photos show the downward-firing machine guns.

Right: AEG J.II in a photo of German aircraft outside the Zeppelin hangar at Trier postwar; it is a detail of the photo below.

Below: Postwar photo of German warplanes gathered outside a Zeppelin hangar at Trier, Germany. Many German aircraft were collected there for distribution to the Allies under terms of the armistice.

Above: An AEG J.II is at right in this postwar lineup, with two LVG C.VI aircraft at left.

Above & Left: A AEG J.II '13' serving with the postwar D.L.R. airline before and after a landing accident. The J.II still wears its wartime camouflage and markings with D.L.R insignia and markings superimposed. Armor was removed from civil AEG J.II aircraft.

Above, Left, & Below: Three different AEG J.II aircraft serving with the postwar D.L.R. airline. *Deutsche Luft-Reederei*, founded in 1917, was the first German airline to use heavier than air aircraft. D.L.R.'s first air services began on 5 February 1919. The company used a stylised crane, designed by Professor Otto Firle, as its logo.

Above: A civil AEG J.II with passenger cabin postwar; unfortunately, the men are not known.

Below: A civil AEG J.II with passenger cabin serving with D.L.R. postwar. In 1919, D.L.R. was one of the founding members of International Air Traffic Association, the predecessor to today's IATA.

In 1923 D.L.R. merged to became part of Aero Lloyd AG, who again merged to became part of Deutsche Luft Hansa in 1926, which also adopted the crane logo.

Above: A group of Turkish officers tours an *Abteilung*; the less than excited aircrew are on the right and their AEG J.I 428/17 forms an interesting background.

Left: A truck retrieves what is left of a downed AEG J-type; given the sprayed camouflage it may be a J.I or early J.II. Although the rest of the aircraft is destroyed, the armored forward fuselage enclosing the engine and cockpits is still intact.

Above: A downed AEG J-type; serial apparently J.2x6/xx. Although resistant to ground fire, the AEG J-types were not as robust as the Junkers J-types.

Above: Likely the wreck of AEG J.II 249/18 brought down in US lines on 27 October 1918.

Above: This view of the inside of the Zeppelin hangar in Trier shows AEG J.II 353/18 tactical '7' with its wings still on. A USAS Salmson 2 A2 is at lower left with a Rumpler C.IV in the center, an LVG C.VI behind the AEG J.II, and Fokker D.VII fighters among the aircraft that can be identified. See page 31 for a closeup of J.II 353/18 without its wings.

Above: Another view of the inside of the Zeppelin hangar in Trier shows AEG J.II 353/18 tactical '7' with its wings on.

AEG J.III

The AEG J.III prototype, designed in September 1918, was finally fitted with the more powerful 260 hp Mercedes D.IVa engine needed from the beginning. *Idflieg* finally accepted what the aircrews knew from the beginning, that more power was needed not only to reduce take-off distance but also to increase speed and improve climb rate and agility. Greater power would improve operational flexibility, overall handling, and stall margin (for flight safety), and the ability to defend against fighter attack.
In addition to the increased power, the J.III had redesigned armor plate that saved 100 kg of weight. As far as is known, the J.III did not reach series production before the Armistice and no photographs have been found.

AEG DJ.I Triplane

Above & Below: The AEG DJ.I triplane was envisioned as a low-altitude fighter armored to withstand ground-fire as it intercepted enemy ground-attack aircraft at low altitude, a role similar to that foreseen for the Sopwith Salamander. It was powered by a 195 hp Benz Bz.IIIb V-8 engine, which did not reach production in time to power any warplane over the front.

Exactly when the idea of an armored, single-seat fighter to oppose Allied ground-attack aircraft was proposed is not known. Surely the concept was another inspiration of *Oberst-Leutnant* Wilhelm Siegert, the commander of *Idflieg*, who throughout the war remained a steadfast proponent of specialized aircraft to fulfill specific tactical assignments. He had no use for general-purpose aircraft. It was reported on 21 October 1917 that AEG was designing an armored fighter, known internally as the AEG PE (*Panzer-Einsitzer*) and may have been later designated DJ.I or DA.I. (In late 1918, two aircraft classifications appeared in various records, both signifying the armored single-seat fighter. They were DJ = D-*Infantrie* and DA = D-Angriff (*attack*). It is not known if these classifications were officially adopted.) The engine, fuel, and pilot were protected by a steel shell forming an integral part of the airframe. The rear fuselage was of aluminum. The triplane wings had AEG's customary steel-tube spars and wooden ribs.

The preliminary trials of the AEG PE triplane, powered by a 195 hp Benz Bz.IIIb V-8 engine, were expected to be concluded in December 1917. From a report dated March 1918, we learn that the AEG *Panzer-Einsitzer* had reached 4000 meters (13,123 ft.) in 48 minutes. The test pilot reported the triplane as easy to fly but not particularly fast. Work on the AEG DJ.I triplane was stopped in May 1918 because of "poor flight characteristics."

AEG DJ.I Biplane

Above: The AEG DJ.I biplane was developed from the earlier DJ.I triplane to improve its performance and flight characteristics. It was powered by the same experimental 195 hp Benz Bz.IIIb V-8 engine, which likely delayed its development. The biplane use the same armored fuselage as the triplane coupled to thick biplane wings unlike those on any other AEG design of the war.

Disappointed by the mediocre flight characteristics and low speed of the AEG DJ.I triplane, AEG engineers brought out a biplane version, the AEG DJ.I fighter, also powered by the 195 hp Benz Bz.IIIb V-8 engine. Sharing the triplane's armored fuselage design, the DJ.I had a wireless wing cellule to minimize damage from ground fire. The construction began in March and was completed in May 1918, but according to *Idflieg* it was not until August that the flight tests began. The reason for the delay is not known but may have been caused by unavailability of the experimental Benz V-8 engine. Converting to a biplane wing cellule improved the aircraft's speed and climb rate. In a 1920 aircraft inventory compiled by *Luftfriko* there were two AEG DJ.I aircraft listed as "property of the factory." Whether this referred to the DJ.I triplane and biplane or the second DJ.I fighter described below has not been determined.

A second AEG DJ.I powered by a 245 hp Maybach Mb.IVa engine was under construction from May to July 1918. In September we learn that the Maybach engine was to be replaced by a new one, pointing to the probability that the second DJ.I had performed some flight trials.

A third AEG DJ.I was reported under construction in September 1918, but there is no record of completion. That three DJ.I prototype fighters existed was consistent with the *Idflieg* policy of ordering one prototype for flight tests, one for static-load testing, and one held in reserve.

Above & Below: The AEG DJ.I biplane was well streamlined and was designed with thicker wings to enable wireless bracing, reducing drag. The 195 hp Benz Bz.IIIb V-8 engine drove a four-bladed propeller.

AEG G.IVk

Above: The AEG G.IVk was a modification of the standard G.IV bomber with armored engine nacelles and forward fuselage crew positions plus a 20mm Becker cannon for the forward and aft gunners for anti-tank duties. Each gunner also had a flexible machine gun to defend against fighter attacks.

Another German armored aircraft type designed for ground-attack was the AEG G.IVk, which was developed from the standard AEG G.IV bomber. AEG G.IV G.1096/16 was armed with a 20mm Becker cannon mounted in an experimental rotating chin turret to investigate the feasibility of attacking tanks and other ground targets from the air. The field of fire was 50 degrees vertical and 90 degrees horizontal. *Idflieg* reported that the aircraft was slated for combat evaluation in December 1917, but further information is lacking. However, ground and air trials demonstrated that the visibility, field of fire, and target acquisition were satisfactory. The cannon installation was reported as "frontline ready", leading to the development of the specialized AEG G.IVk ground-attack aircraft.

As a result of the progress made with G.1096/16, in March 1918 *Idflieg* ordered five AEG G.IVk grount-attack aircraft for service trials. Numbered G.500/18 to G.504/18, they were armed with two 20mm Becker cannons, one mounted in the lower nose turret and a second in a rear turret that allowed for firing over the fuselage or downward through a 'Gotha-type' tunnel. For defense against fighters, two Parabellum machine guns were mounted on LVG gun rings (copies of the Scarff) in the nose and rear turrets. Like the rear cannon, the rear Parabellum could also fire downward through the fuselage tunnel. Four 50 kg P.u.W. bombs could be carried under the fuselage. Four vertical magazines in the front cockpit and three in the rear held 12.5 kg P.u.W. fragmentation bombs. Chrome-nickel steel armor plating 5 mm thick covered the floor and sides of the fuselage, extending from the nose around the rear gunner's cockpit. The engine nacelles were armored as well.

Flight trials of the prototype, G.500/18, in February 1918 demonstrated good results but the biplane tail and armor plating required minor changes. The cannon firing trials performed in April 1918 were satisfactory. Though dispatched to the front in October–November 1918, it is not known if the G.IVk saw operational service. As part of the Armistice stipulations, the Germans turned over the four production AEG G.IVk ground-attack aircraft (G.501/18–G.504/18) to the British in January 1919.

This Page: Four production AEG G.IVk ground-attack aircraft, G.501/18–G.504/18, were turned over to the British as part of the armistice terms; G.503/18 appears to the aircraft shown here.

Above: The prototype AEG G.IVk was G.500/18 shown here on the AEG factory airfield at Henningsdorf. The four production aircraft had a different tail design; see below for comparison. G.500/18 wears an interesting camouflage of two-color sprayed camouflage on the fuselage, dark engine nacelles, and printed camouflage fabric on the fins and rudders. The 20mm Becker cannon in the nose mounting is prominent. This aircraft was dispatched to the front for operational use in October–November 1918, but it is not known if it was used in combat.

Above: The AEG G.IVk G.503/18 was one of the four production G.IVk ground-attack aircraft turned over to the British according to the terms of the armistice. The vertical tail surfaces of the four production machines differed from the prototype. The aft gunner could fire his 20mm Becker cannon conventionally to the sides or down through the fuselage in a Gotha-like tunnel, and could do the same with his machine gun. The forward gunner had a forward-firing 20mm Becker cannon for anti-tank duties and a flexible machine gun to defend against fighter attack. No conventional bomb bay was fitted but the AEG G.IVk had four vertical tubes about two feet long mounted in the front cockpit and three in the rear cockpit. These tubes were perhaps intended for launching flares or special anti-tank bombs that *Idflieg* had developed.

Above: The nose turret of the AEG G.IVk showing the 20mm Becker cannon. Armored doors to close the mounting are at the sides of the opening for the cannon.

Above: The rear gunner of the AEG G.IVk in his turret handling his 20mm Becker cannon. Although no bomb bay was fitted and supposedly no bomb racks, bombs can be seen underneath his position. He could fire both his cannon and machine gun to the sides and downward through a Gotha-style tunnel through the fuselage.

Above: The nose turret of the AEG G.IVk with the 20mm Becker cannon seen from the inside. Although the cannon pointed forward and down for ground-attack, the gunner's machine gun was in a conventional gun ring to defend against fighters.

Above: An Australian soldier pokes his head our of the nose turret of the AEG G.IVk; another stands in the cockpit. The armor plate was difficult to form, so all armor on the aircraft was either a flat plate or a simple curve. Compound curves were simply too difficult to form at the time.

AEG G.IVk G.500/18

AEG G.IVk G.501/18

AEG G.IVk G.503/18

AEG Bombers

Despite the wide variety of AEG aircraft that served operationally, AEG is perhaps best known for its twin-engine bombers. These originated from development of a 'battle plane' that was essentially an aerial cruiser armed with flexible machine guns and bombs. This aircraft lead to a series of designs that became twin-engine bombers based on operational experience.

In addition to these G-type bombers, AEG also built the successful N.I single-engine night bomber and a giant bomber, the R.I, that suffered a fatal crash during flight testing.

Right: An aviator poses with his AEG G.II. The AEG G-types all followed the same basic configuration; conventional biplanes with steel-tube structures covered by fabric and two engines mounted as tractors. Nearly all other German bombers had engines mounted as pushers.

AEG G-Type Specifications

	G.I	G.II	G.III	G.IV	G.V
Engine	2x100 hp Mercedes D.I	2x150 hp Benz Bz.III	2x220 hp Mercedes D.IV	2x260 hp Mercedes D.IVa	2x260 hp Mercedes D.IVa
Span Upper	16.00 m	16.20 m	18.44 m	18.40 m	27.24 m
Span Lower	15.20 m	15.20 m	17.20 m	17.40 m	26.30 m
Chord Upper	2.20 m	2.20 m	2.50 m	2.40 m	2.80 m
Chord Lower	2.20 m	2.20 m	2.50 m	2.40 m	2.39 m
Gap	2.30 m	2.30 m	2.60 m	2.20 m	3.00 m
Wing Area	61.0 m^2	61.0 m^2	74.0 m^2	??? 68.7 m^2	127.2 m^2
Length	8.7 m	9.1 m	9.20 m	9.70 m	10.80 m
Track	3.15 m	3.15 m	2.85 m	5.10 m	4.85 m
Empty Weight	1,160 kg	1,450 kg	1,940 kg	2,400 kg	2,700 kg
Loaded Weight	1,610 kg	2,050 kg	2,560 kg	3,635 kg	4,800 kg
Maximum Speed	125 kmh	140 kmh	150 kmh	165 kmh	145 kmh
Climb, 1000m	—	11 min.	6 min.	5 min.	6 min.
Climb, 2000m	—	—	—	11 min.	12 min.
Climb, 3000m	—	—	—	21 min.	23 min.
Climb, 4000m	—	—	—	40 min.	34 min.
Armament	2 flexible machine guns, small bombs	2 flexible machine guns, 200 kg bombs	2 flexible machine guns, 240 kg bombs	2 flexible machine guns, 300 kg bombs	2 flexible machine guns, 1,000 kg bombs

Note: The AEG G.IVb wing span was enlarged to 24 m; this enabled a 1,000 kg bomb to be carried. For short missions up to 1,500 kg of bombs could be carried.

AEG K.I

In March 1914 the German general staff sanctioned the development of the *Kampfflugzeug* (battle plane), and in July 1914 *Idflieg* issued specifications for the type in preparation for a competition planned for spring 1915. The *Kampfflugzeug* idea was basically an 'aerial cruiser' armed with machine guns and bombs. The aircraft was to have 200 hp, carry a crew of three, and have an endurance of six hours.

AEG responded to the requirement with a biplane powered by two 100 hp Mercedes D.I engines; the factory designation was GZ1 and the military designation was AEG K.I. The K-type designation was soon changed to G-type, the 'G' standing for '*Grossflugzeug*' or large aircraft, later to become synonymous with twin-engine bombing aircraft. The K.I had side-by-side seating for two crewmen and a single flexible machine gun was mounted in a nose turret because the aircraft was intended to chase and destroy enemy airplanes. The airframe was constructed of self-fused (autogenous) welded steel tubes. The nose was covered with light armored plate and the rest of the aircraft was fabric covered.

In January 1915 the AEG K.I prototype was first flown by test pilot Willy Kanitz and gave promising results during flight tests in January–February. This convinced *Idflieg* to order a second prototype as the AEG G.I for combat evaluation.

Above: The AEG K.I designed to the flawed *Kampfflugzeug* concept was the first twin-engine AEG design. All subsequent AEG G-types followed the same basic configuration; conventional biplanes with fabric-covered steel-tube structures with two engines mounted as tractors. Almost all other German bombers had engines mounted as pushers. The K.I was a compact design with good handling qualities; only one K.I was built.

AEG G.I

The second AEG *Kampfflugzeug* prototype, now designated AEG G.I but internally retaining the company designation GZ1, was completed in March 1915 and may have incorporated components from the K.I. The G.I differed from the K.I primarily in its crew and armament; the G.I had three crewmen and flexible machine guns in fore and aft turrets. Power remained two 100 hp Mercedes D.I engines

The AEG G.I was shipped from the factory on April 24, 1915 for the 4.*Armee* for combat assessment without first being tested at Adlershof. Only one G.I was built.

Above: The AEG G.I differed from the K.I primarily by having an additional crewman with a flexible gun mounted aft. The G.I may have incorporated some components of the K.I.

Right: The nose of the AEG G.I opened to show details of the front gunner's cockpit.

AEG G.II

Prior to the operational trials of the AEG G.I, *Idflieg* ordered six aircraft of an improved type, the AEG G.II (factory designation GZ2). *Idflieg's* requirements for the AEG G.II included two 150 hp Benz Bz.III engines, a crew of two with three seats, a single machine gun, a 200 kg bomb load, and 150 kg of armor plate (front and side 2.3 mm and floor 1.5 mm). Controlling a multi-engine aircraft after the failure of one engine was dependent on both pilot technique and design of the aircraft; accordingly *Idflieg* urged AEG "to make every effort to assure that the aircraft would maintain a straight flight path with only one engine running at full power." With the more powerful engines in the G.II a larger fin and rudder were needed to main control with asymmetric thrust.

As expected the more powerful engines gave the AEG G.II better performance and greater load-carrying capability, and *Idflieg* ordered a second batch of 12 G.II aircraft on May 6, 1915 before the results of the combat trials of the AEG G.I were known. The first two G.II aircraft were completed in May and reached the front in June, with a maximum of 13 at the front in December.

The AEG G.II was not only a new aircraft but was exploring a new combat role as a multi-engine *Kampfflugzeug*, or aerial cruiser. The G.II was used both to escort single-engine reconnaissance and bomber aircraft and to attack enemy aircraft.

Based on operational experience the first six G.II aircraft (G.2–7/15) were extensively modified for both technical and operational reasons, with the result that no two aircraft were alike. The square, armor-plated nose was replaced by a streamlined, unarmored one. Two gravity tanks and new oil tanks were installed. Because the small fin and rudder failed to provide adequate directional control with one engine out, some machines were retro-fitted with triple rudders, becoming standard beginning with the second G.II production batch (G.19–30/15).

Despite all the modifications to the G.II, including adding a second and sometimes a third machine gun, operational experience soon made it clear that the multi-engine *Kampfflugzeug* concept was a failure. The *Kampfflugzeug* had only modest success as a multi-seat escort and was too slow and cumbersome to catch faster, more maneuverable enemy aircraft – and most enemy single-seaters were faster and more maneuverable.

However, bomb racks were installed in all G.II aircraft and aircrews soon discovered that bombing was the most effective role for the G.II. By mid-1916 *Idflieg* summarized the operational record of the G.II saying it "had fared poorly in air combat, but had been successfully employed as a bomber in squadron strength." The AEG G.II thus discovered the true role of the G-type as a bomber. It remained at the front through June 1917 and set the standard for future AEG bombers. A total of 27 G.II aircraft were delivered before production was shifted to the improved G.III in May 1916.

Below: AEG G.II G.3/15 was the second production G.II and has the single fin and rudder originally used.

Above: An AEG G.II in flight. This image has also been identified as a G.IV, but the engines and radiators indicate it is a G.II.

Left: AEG G.II G.7/15 with armored nose.

Below: AEG G.II G.19/15 serving with *Flieger-Abteilung* 22 features triple rudders and an unarmored nose. It appears to have a two-color sprayed camouflage.

Above: Under-fuselage bomb racks are visible on this AEG G.II without nose armor.

Below: Future ace Rudolph Berthold flew AEG G.II G.21/15 while serving with *FFA* 23.

Above: Rudolph Berthold sits in the cockpit of AEG G.II G.26/15 that he also flew while serving with *FFA* 23.

Below: Rudolph Berthold standing in the cockpit AEG G.II G.21/15 during a royal visit by Duke Ernst August of Brunswick and Prince August Wilhelm of Prussia to *FFA* 23 on Oct. 23, 1915.

Above: Aircrew of *Flieger-Abteilung* 6 pose with an AEG G.II of that unit.

Left: Closeup of an AEG G.II shows the multitude of drag-producing struts and bracing wires.

Right: AEG G.II after a landing accident.

Above: Aircrew pose with an AEG G.II with armored nose.

Above: Groundcrew move an AEG G.II with unarmored nose.

Above: Aircrew pose with an AEG G.II with armored nose.

Below: Front view of an AEG G.II without nose armor.

Above: Aircrew and groundcrew pose with an AEG G.II with armored nose.

Above: An AEG G.II without nose armor rests on an airfield with a Fokker E-type in the foreground.

Above: AEG G.II with triple rudders and two-color sprayed camouflage finish.

Below: This AEG G.II is at the factory and has outsize triple rudders fitted without vertical fins. This was a prototype; as seen above the production aircraft with triple rudders had a central fixed fin.

Above: AEG G.II with triple rudders and early single color finish.

Below: AEG G.II with triple-rudders and two-color sprayed camouflage finish.

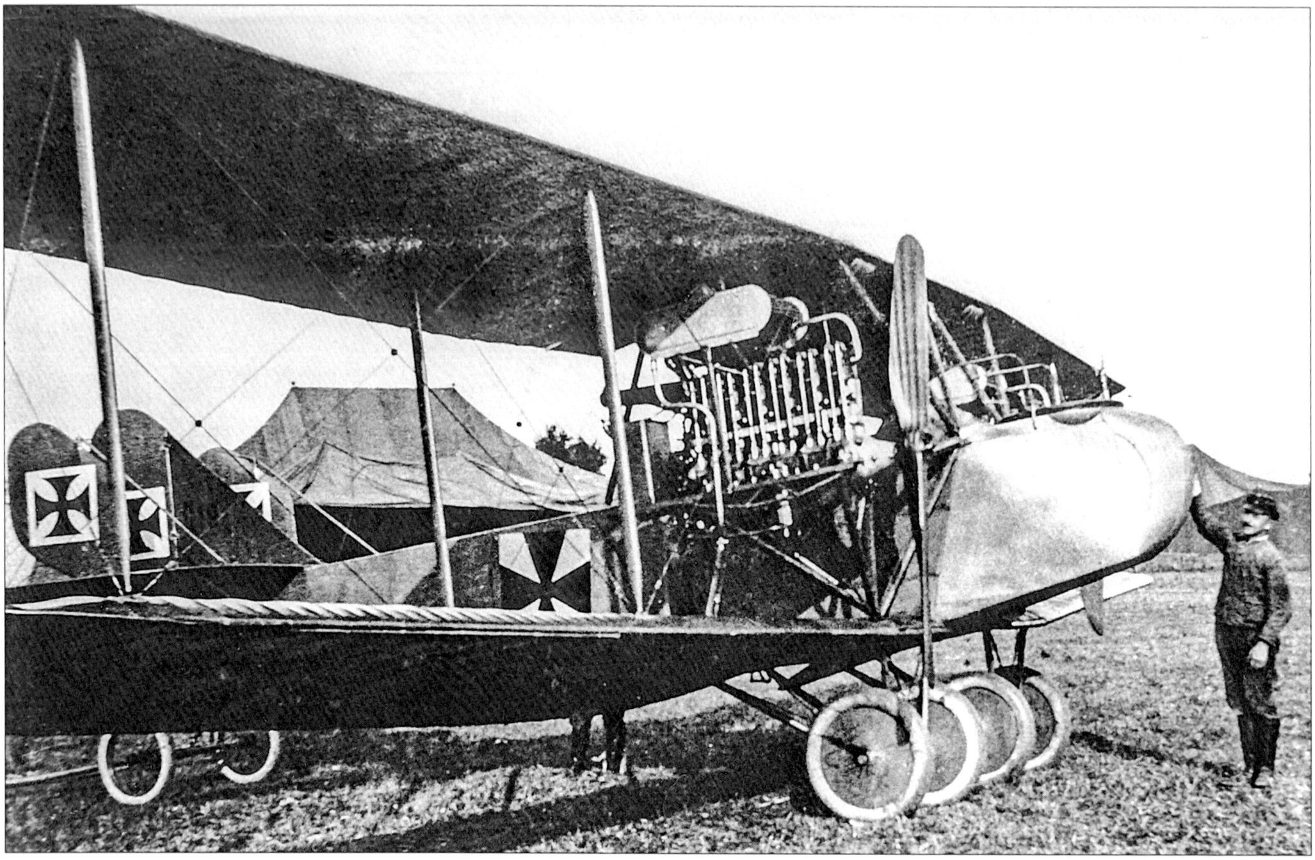

Above & Below: AEG G.II with triple rudders and fairly dark single color finish. There is a lot of contrast between the overall finish and the white background of the national insignia, especially on the rudders.

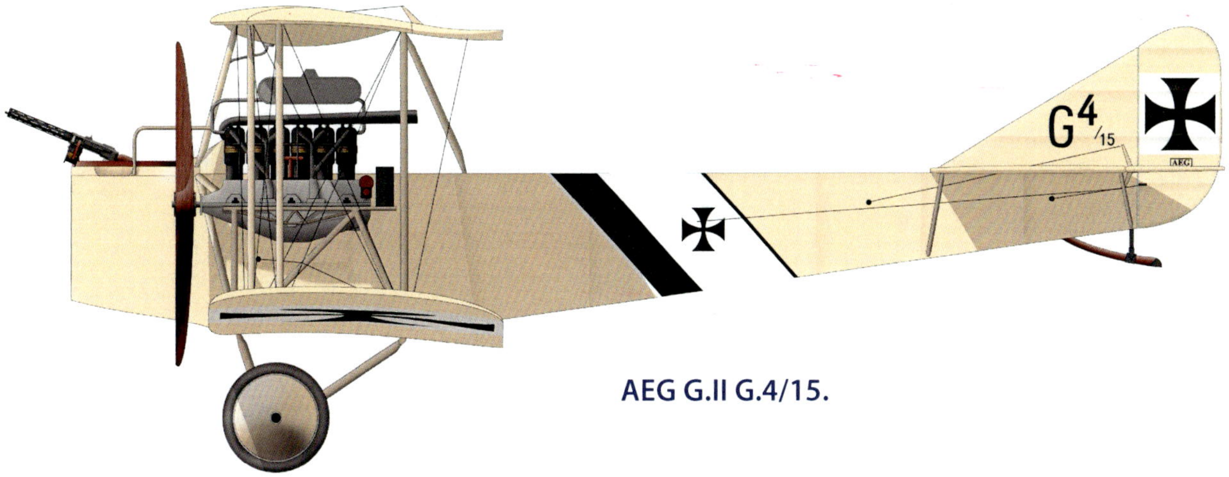

AEG G.II G.4/15.

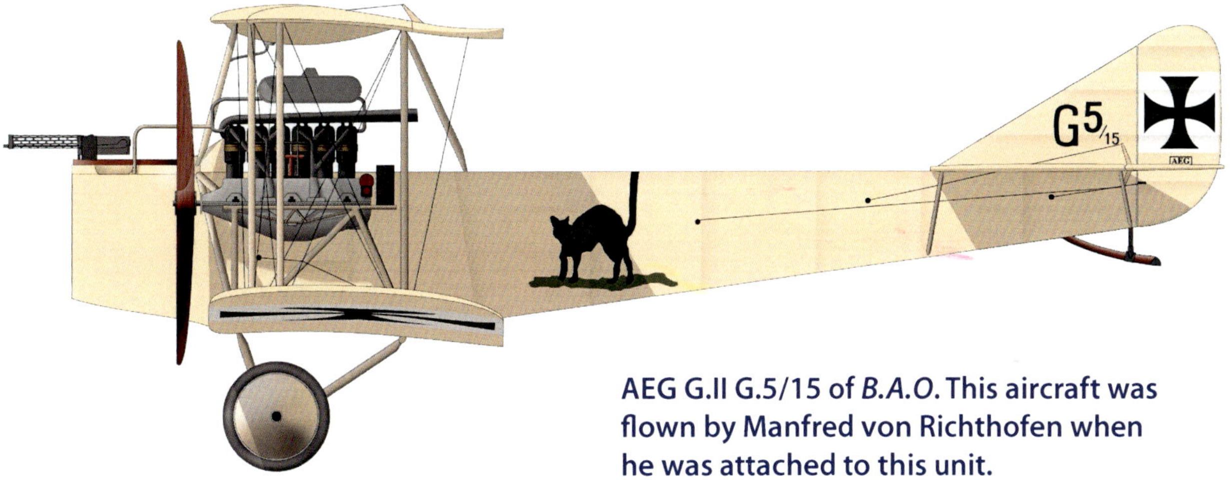

AEG G.II G.5/15 of *B.A.O.* This aircraft was flown by Manfred von Richthofen when he was attached to this unit.

AEG G.II G.7/15 of *FFA* 42 at Strasbourg in 1916.

AEG G.II G.19/15 possibly flown by Walter von Bülow at *FFA*22.

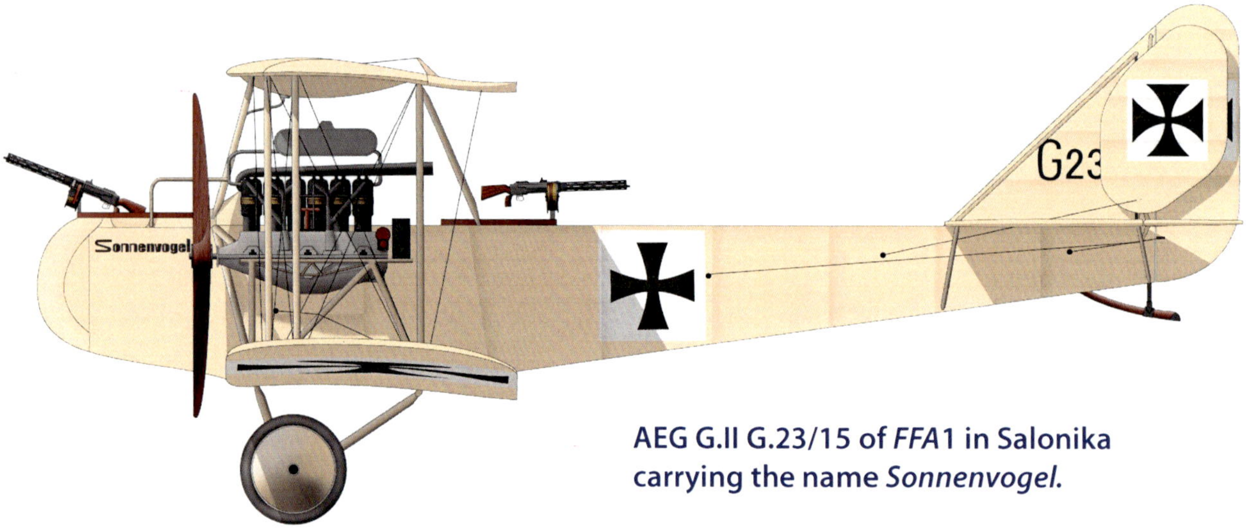

AEG G.II G.23/15 of *FFA*1 in Salonika carrying the name *Sonnenvogel.*

AEG G.II of *FA* Karlhorst. The text on the fuselage translates as: Training Aircraft 5 of the Aircraft Detachment Karlhorst of the Aerial Photography Command.

Above: AEG G.II with triple rudders and single color finish.

Below: AEG G.II with triple-rudders and single color finish. The color appears darker than some of the earlier monotone finishes used on G.II aircraft like that shown above; on this aircraft there is more contrast between the white background for the national insignia and the overall color. Unfortunately, the serial number is not visible.

Above & Below: AEG G.II with single rudder. The single color finish is so light there is almost no contrast between the overall finish and the white background of the national insignia.

AEG G.III

When *Idflieg* ordered the first six G.II on April 1, 1915, AEG was also requested to build a *Kampfflugzeug* with two 220/240 hp engines, the type to be decided later. The intention was to provide greater performance and payload, and the resulting aircraft was designated the AEG G.III (factory designation GZ3). A crew of three, 200 kg of armor protection, and a bomb load of 240 kg were specified. Two machine guns or a cannon mounted in the nose and a rear machine gun was the specified armament.

The G.III was very similar to the G.II although the wingspan was 2.24 m longer. Two of the new, 220 hp Mercedes D.IV straight-eight engines were fitted because these were the most powerful engines then in production. A four-bladed propeller was used to absorb all the power of the engine, but the long crankshaft was subject to fractures in service. The AEG specification chart shows December 1915 as the date the first G.III was completed, but the first three production aircraft were not delivered until May 1916. The G.III prototype was tested with a single fin and rudder and an early production aircraft (G.53/15) was delivered with a triple rudder for comparison. All other production G.III bombers had the single fin and rudder and most had external bomb racks.

The G.III reached the front in June 1916 and was first used as an escort aircraft for single-engine bombers, but it quickly became clear that bombing was the appropriate role for the G.III. *Kampfgeschwader* 1 became "the first formation to be completely equipped with twin-engine aircraft of the G-category for the sole purpose of bombing." Serving with *KG*1 in Macedonia, the G.III was primarily used as a bomber but at least two strafing attacks were made. The last three G.III bombers were delivered in January 1917, bringing total G.III production to 45 aircraft; the AEG G.IV then followed the AEG G.III in production.

Above: AEG G.III G.54/15 serving at the front. The four-blade propellers are a key G.III identification feature.

Below: AEG G.III G.52/15 serving at the front. The G.III was basically an enlarged, more powerful G.II.

Above: AEG G.III G.233/15; the enlarged rudder with aerodynamic balance helped the pilot maintian control with an engine out despite its more powerful engines. The aircraft wears a very light overall monotone finish.

Below: This portrait of an AEG G.III became Sanke Card 1060. The finish was two camouflage colors sprayed on.

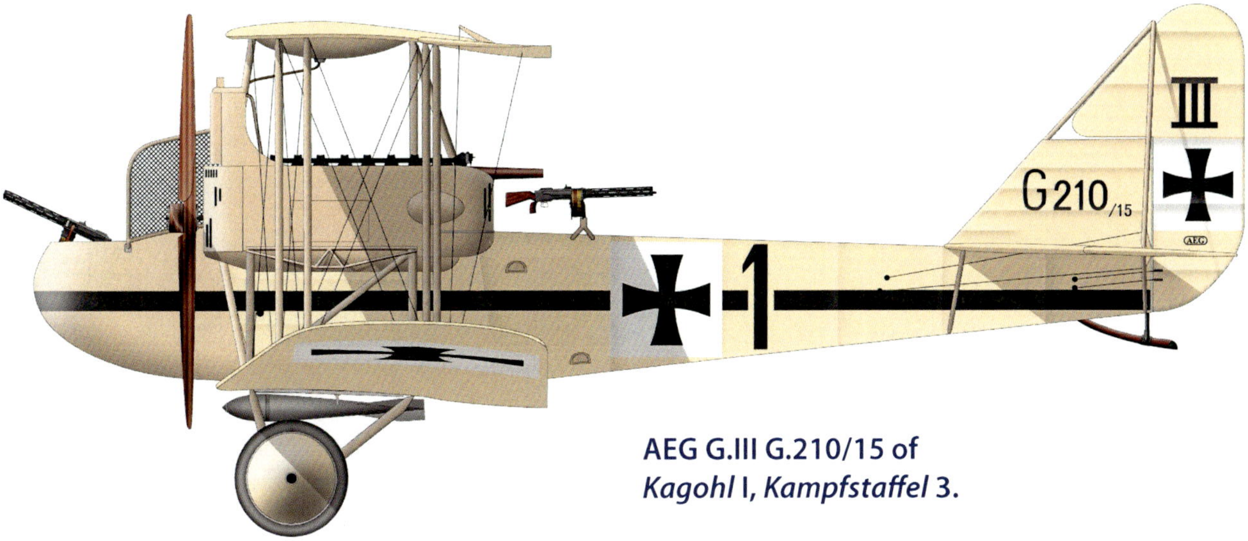

AEG G.III G.210/15 of
Kagohl I, *Kampfstaffel* 3.

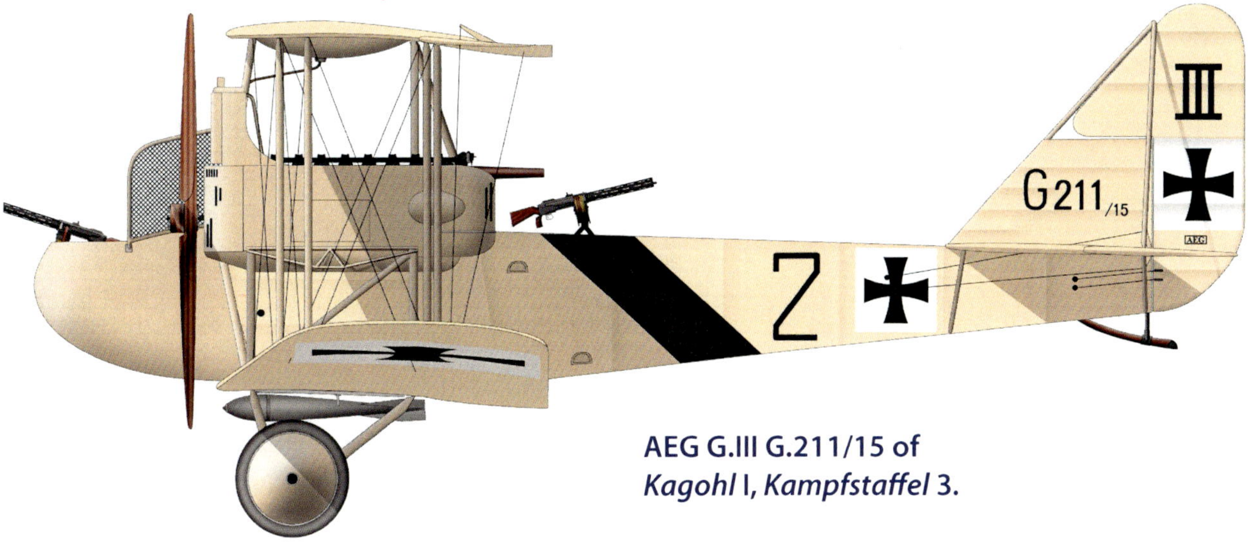

AEG G.III G.211/15 of
Kagohl I, *Kampfstaffel* 3.

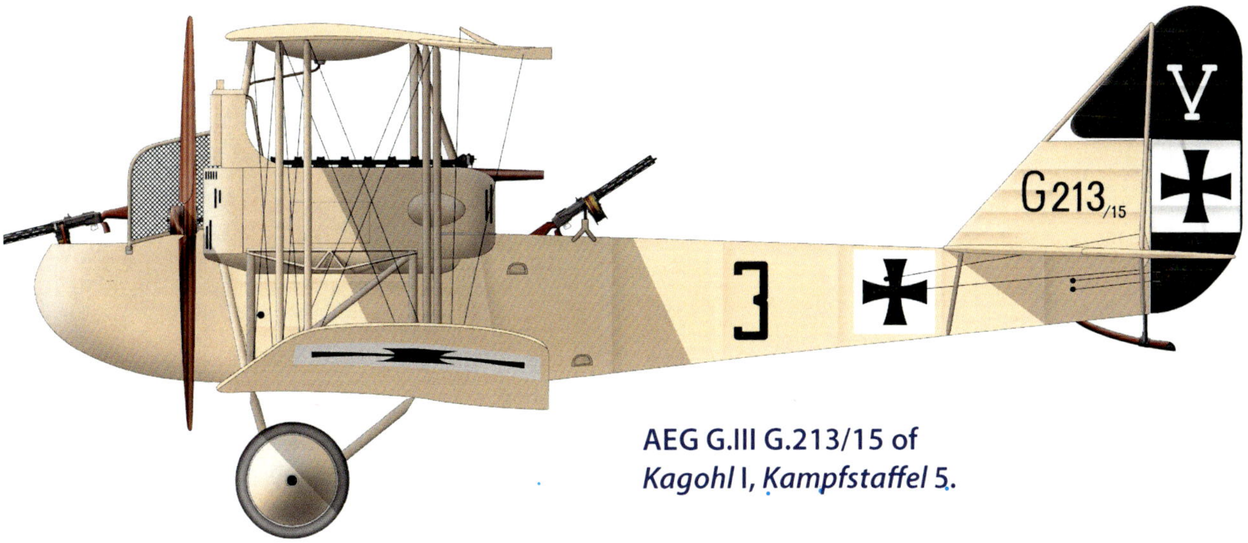

AEG G.III G.213/15 of
Kagohl I, *Kampfstaffel* 5.

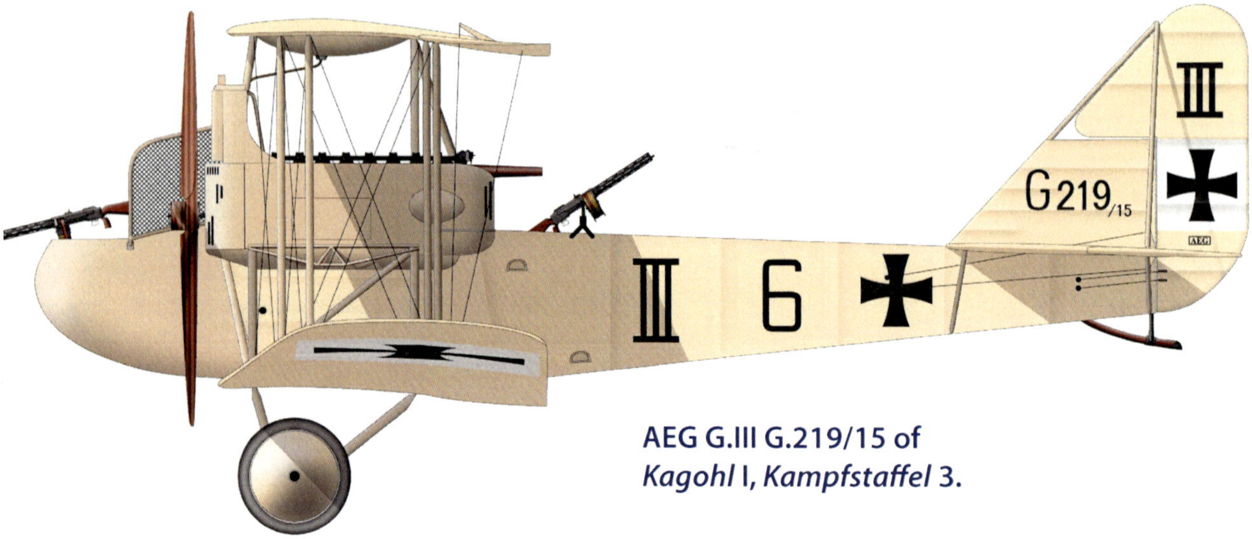

AEG G.III G.219/15 of
Kagohl I, *Kampfstaffel* 3.

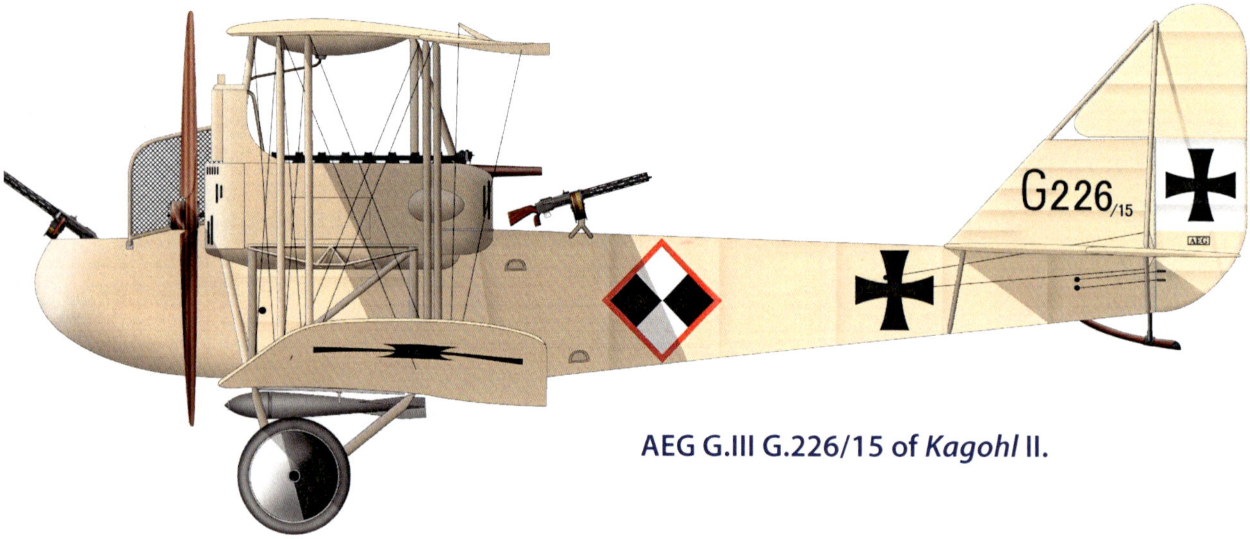

AEG G.III G.226/15 of *Kagohl* II.

AEG G.III G.227/15.

Above: AEG G.III G.213/15 tactical number '3' serving with *Kagohl* I, *Kasta* 5 based on the Roman numeral on the rudder. The straight-eight cylinder Mercedes had good power but the long crankshaft was subject to failure, especially in multi-engine aircraft.

Below: AEG G.III G.216/15 in overall light finish being repositioned on the airfield.

Above & Facing Page: Closeup of an AEG G.III being serviced. The pilot's headrest is unusual for a twin-engine aircraft. The photo at right was torn across the bottom.

Below: The four-blade propellers were a trademark of the AEG G.III with its 220 hp Mercedes D.IV engines.

Above: A mix of light-colored and camouflaged AEG G.III bombers of a *Kampfgeschwader* are lined up.

Above: The aircrew of an AEG G.III bomber flank damage to their aircraft likely caused by anti-aircraft fire. This photo gives a good view of the complex struts supporting the 220 hp Mercedes D.IV engines.

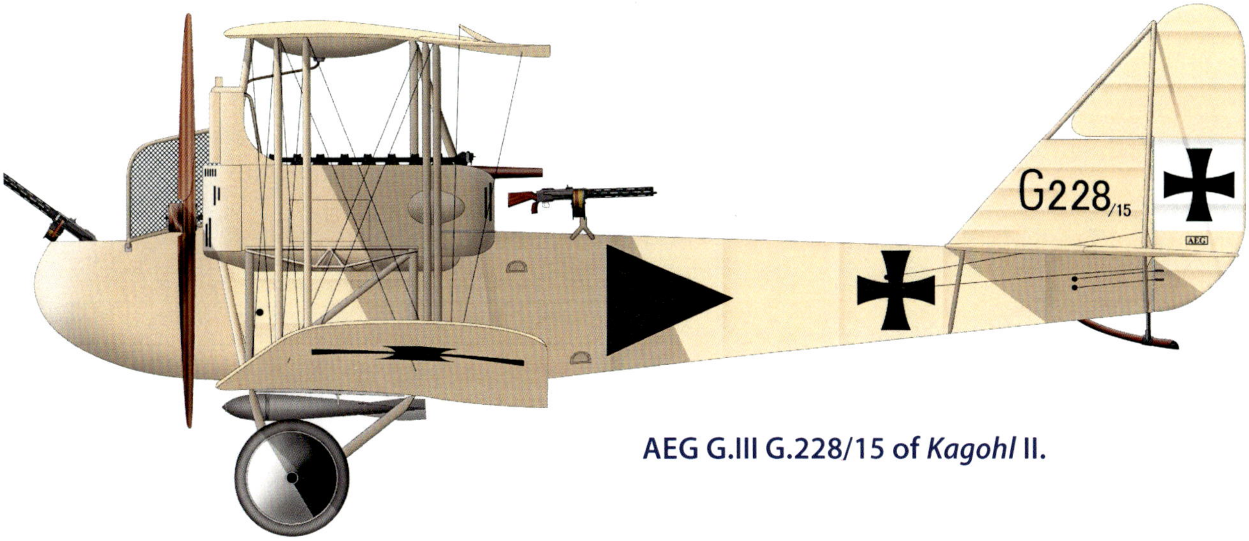

AEG G.III G.228/15 of *Kagohl* II.

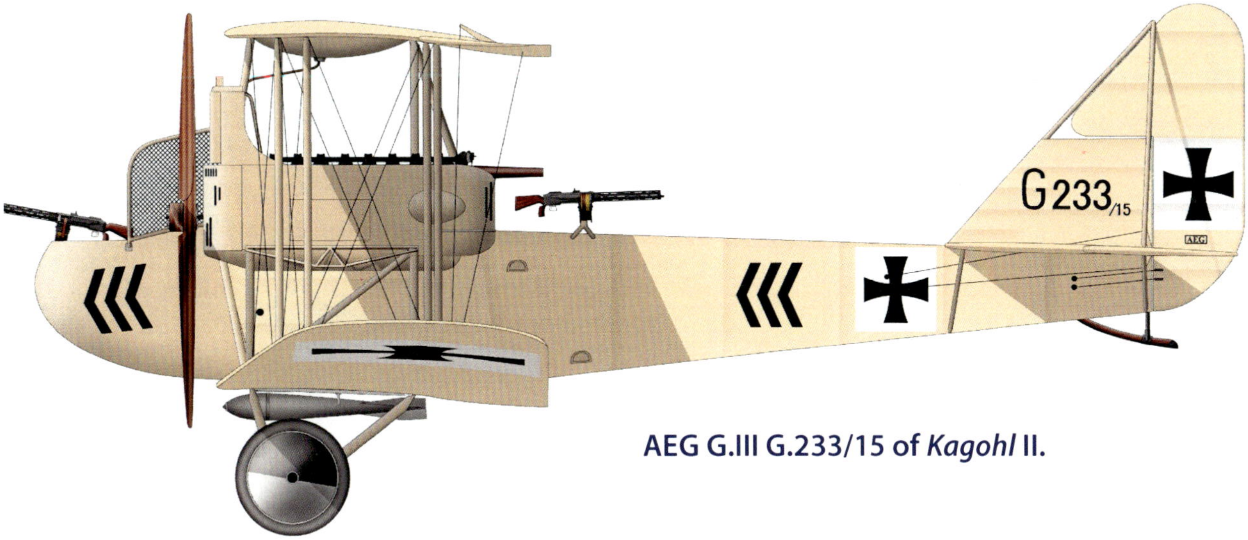

AEG G.III G.233/15 of *Kagohl* II.

AEG G.II G.235/15 seen at the *Geschwaderschule* at Paderborn. Most of the original plain finish has been recovered with night lozenge; the left wheel is uncovered but the right wheel is covered.

Left & Below: Twin-engine aircraft were also subject to landing accidents. A gust of wind at the wrong time or poor pilot technique could result in touching down in a sideslip, often with dramatic results.

Below: AEG bomber dismantled for transportation by rail. The railroads were more reliable than the aircraft of the day.

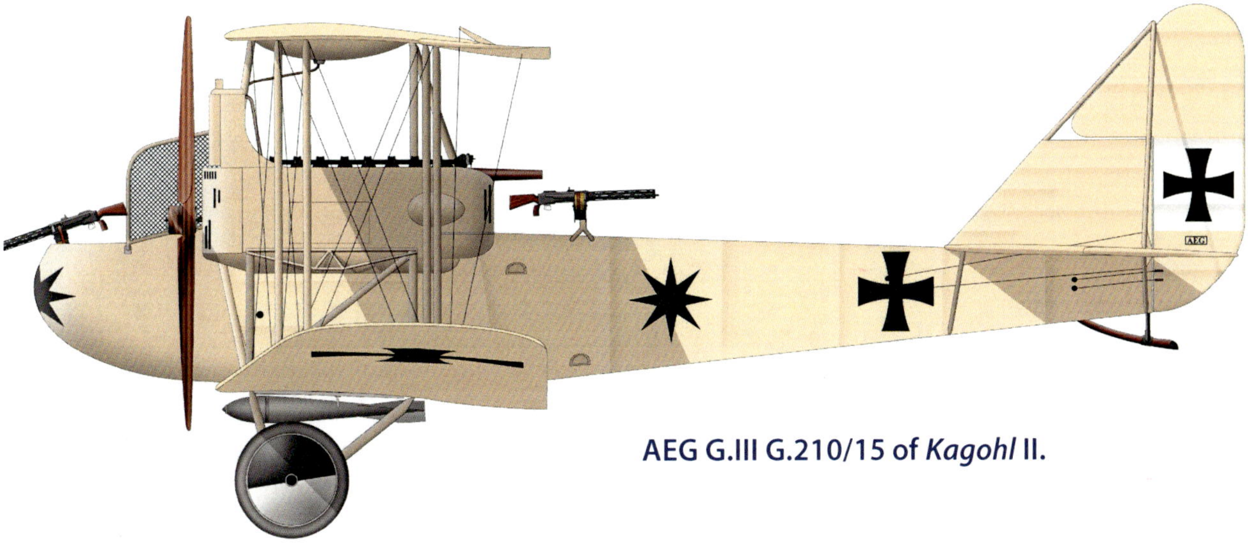

AEG G.III G.210/15 of *Kagohl* II.

AEG G.III G.143/16 seen at the *Geschwaderschule* at Paderborn. Most of the original plain finish has been recovered with night lozenge.

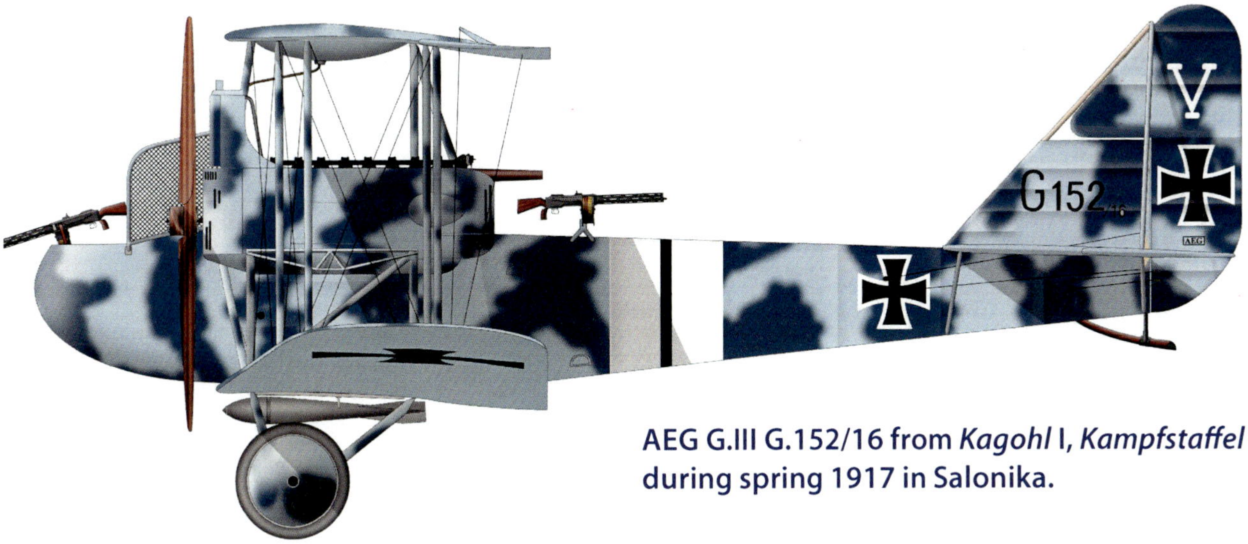

AEG G.III G.152/16 from *Kagohl* I, *Kampfstaffel* 5, during spring 1917 in Salonika.

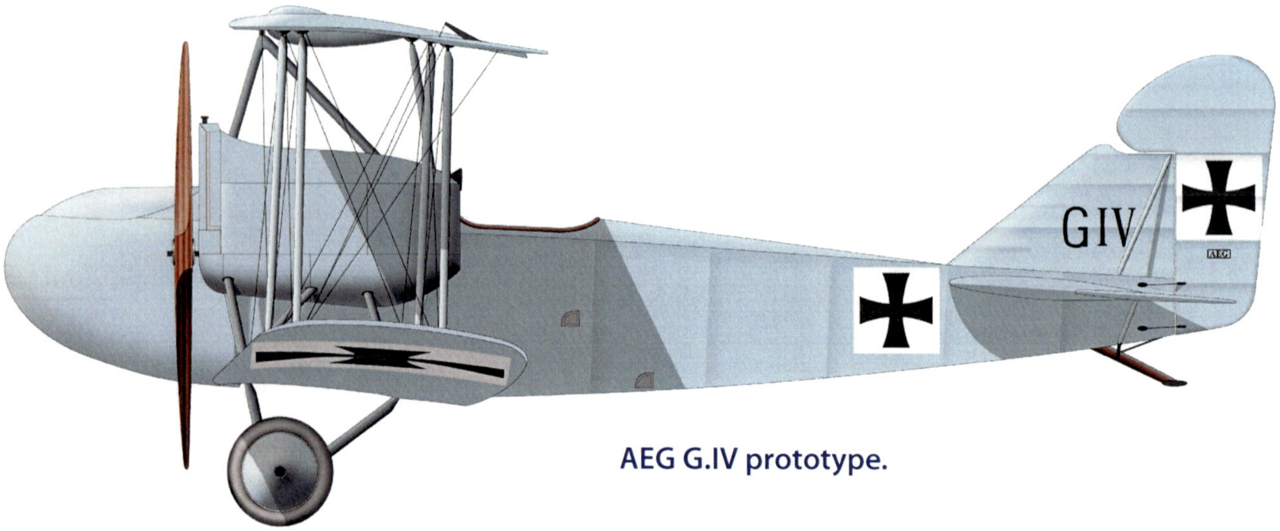

AEG G.IV prototype.

AEG G.IV G.155/16, the 3rd G.IV built.

AEG G.IVb G.168/16.

AEG G.IV

The next step in development of the AEG twin-engine bombers was to give the aircraft more powerful and reliable engines, and the G.IV used the new 260 hp Mercedes D.IVa. The new engine not only had more power but was a six-cylinder engine with more robust crankshaft and was notably more reliable than the earlier D.IV straight-eight.

In March 1916 *Idflieg* placed an initial order for 40 AEG G.IV bombers; the requirements specified a load of 25 x 12 kg or 6 x 50 kg bombs (300 kg total), four machine guns or cannon, a climb to 5,000 meters in 75 minutes, and a top speed of 140 km/h. The G.IV (factory designation GZ4) prototype was completed in September 1916 and exceeded all the requirements except for climb.

The first G.IV production aircraft were delivered in January 1917, immediately after completion of the last G.III aircraft. The G.IV was essentially the same configuration but airframe dimensions differed and the more complex joints were machined from billet steel for greater strength. Its tough steel tube airframe made the AEG G.IV much more robust then the wooden Gotha and Friedrichshafen bombers, especially in crashes. Furthermore, the AEG G.IV was easier to fly than those bombers and did not require the nose-mounted or wing-mounted auxiliary landing gear to prevent nose-overs on landing. For its robustness and better handling the AEG G.IV earned a better reputation amongst German bomber aircrew than its competitors.

Starting in April 1917 the G.IV began to replace the G.III at the front, and served until the end of the war.

AEG G.IV Engine Experiments

Unlike nearly all other German bombers, the engines of AEG bombers were mounted in tractor configuration, not as pushers. AEG experimented with other engines in the G.IV between September 1917 and March 1918, including the 245 hp Maybach Mb.IVa and 300 hp Basse & Selve BuS.IVa engines, to increase the climb rate and payload. Both engines gave better performance than the standard 260 hp Mercedes D.IVa, but the Mercedes remained the G.IV production engine because other aircraft types had priority on the new engines.

In parallel with the experiments with different engine types, experimental turbo-charged Mercedes engines were installed in a G.IV for flight tests. Unfortunately, that aircraft was destroyed in March 1918. Initially Schwade compressors were used, but manufacturing problems at Schwade motivated AEG to build experimental turbo-compressors that were tested in the summer of 1918. In September 1918 *Idflieg* reported that a G.IV bomber equipped with AEG turbo-compressors raised the operational ceiling from 4200 meters to 6000 meters. As a result AEG received a contract to supply 20 turbo-compressors for combat evaluation in bomber aircraft.

Below: Three-bay AEG G.IVb G.168/16 was from the first G.IV production batch and was rebuilt as an extended-span G.IVb.

AEG G.IVa

The AEG G.IVa has not been identified in *Fliegertruppe* records, and it is unknown if it was an un-built project or an actual variant of the G.IV.

AEG G.IVb

In mid-1917 some AEG G.IV bombers from the first production batch were given wings of increased span and area to improve their climb rate and ceiling. These aircraft, fitted with three-bay wings of 24.0 m span, were designated the G.IVb. In September 1917 *Idflieg* decreed that "since the AEG G-types are (now) employed solely for night bombing it is possible to forego the higher climb rate of the three-bay machine in favor of the superior speed and maneuverability of the twin-bay machine. Henceforth only twin-bay aircraft will be built." Given the extra load carrying capability, some G.IVb bombers were modified to carry the 1,000 kg (2,200 lb) P.u.W. bomb.

AEG G.IVb-Lang

In March 1918 an AEG G.IVb powered by two 300 hp Basse & Selve BuS.IVa engines and fitted with a lengthened fuselage and a box tail to improve engine-out control was flight tested with good results. The box tail had two rudders and biplane horizontal stabilizers and elevators to give the aircraft better controllability during engine-out operations, and succeeded to the extent that the aircraft could even be turned toward the running engine. The modified aircraft, designated AEG G.IVb-lang, (*lang* = long, for the lengthened fuselage) was the forerunner of the AEG G.V.

On 30 July 1919, AEG test pilot Paul Schwandt and seven passengers (1,000 kg useful load) broke the official world record by reaching 6,100 meters in the AEG G.IVb-lang 856/17. At the time, the record machine was powered by two 260 hp Mercedes D.IVa engines supercharged by two AEG turbo-compressors driven by shafts from the rear of the engines.

Above & Below: AEG G.IV G.1125/16 from the second G.IV production batch is shown here in French markings after being brought down on December 23, 1917 by anti-aircraft fire. It wears AEG-style lozenge night-bomber camouflage.

Above: AEG G.IV G.581/17 from the third G.IV production batch wore the AEG lozenge night-bomber camouflage and intermediate-style thick version of the straight-sided *Balkenkreuz* national insignia.

Below: AEG G.IV G.1131/16 from the second G.IV production batch is being evaluated with other captured German aircraft, including an LVG C.V and Albatros D.Va, at the French aviation test center at Villacoublay.

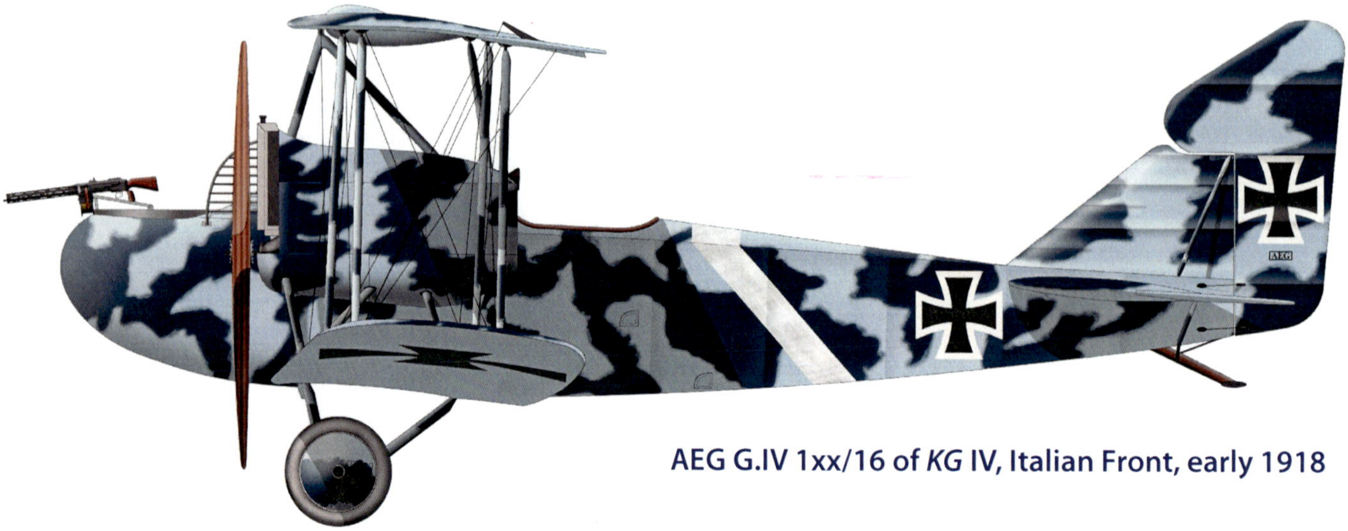

AEG G.IV 1xx/16 of *KG* IV, Italian Front, early 1918

AEG G.IV G.1131/16, unit unknown, June 1918.

AEG G.IV G.1131/16, after capture by the French and repainted in French markings, photographed at Villacoublay on 21 June 1918.

Above: AEG G.IV G.1131/16, slightly bent after its hard forced landing at night, at Villacoublay shows details of its engine installation, camouflage, and bomb racks. Storage for signal flares is provided on the outside of the front gun position.

Below: AEG G.IV G.1256/18 from the last G.IV production batch wearing late-style national insignia. It was turned over to the British in late December 1918 in accordance with Armistice requirements.

AEG G.IV G.567/18 of *Staffel* 27, *Bogohl* 8b. The "27" on the fin indicates *Staffel* 27; "7" is its tactical number within the Staffel.

AEG G.IV G.848/17, *Bogohl* VIII, *Bosta* 27, perhaps the aircraft of *Oblt*. Fritz Diemer, Maria Alter Aerodrome, May 1918.

AEG G.IV G.572/17 brought down behind American lines on 2 June 1918.

Above: AEG G.IV 1125/16 of *Bogohl* III/15 was brought down by anti-aircraft fire on Dec. 23, 1917. The *Eiserneskreuz* on the rudder has already been 'souvenired'. It carries the typical AEG lozenge night bomber camouflage. The crew became PoWs.

Below: AEG G.IV 155/16 is the third production machine during flight evaluation at Adlershof. It has the early two-color sprayed camouflage AEG used before changing over to the lozenge night-bomber camouflage as seen above. The underwing bomb racks are visible; the hole under the nose is for dropping 12.5 kg P.u.W. bombs by hand.

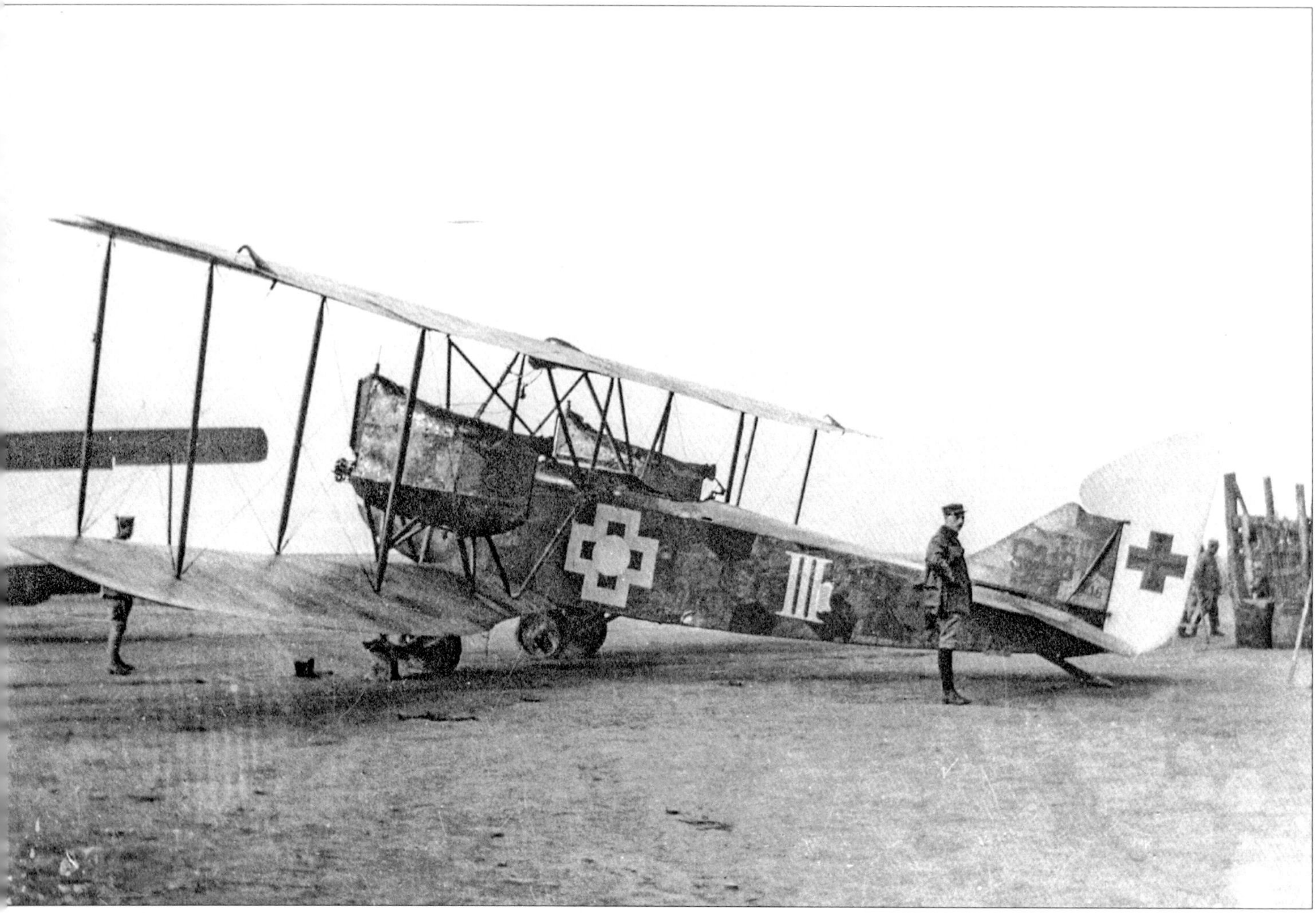

Above: This AEG G.IV was captured April 24, 1918. It carries the typical AEG lozenge night bomber camouflage and interesting markings. It has the large, pointed rudder installed on all production G.IV aircraft.

Below: The ground crew gathers around this AEG G.IV possibly in preparation for moving the aircraft.

Right & Below: More images of AEG G.IV G.1256/18 from the last G.IV production batch. These were taken postwar and the German national insignia have been painted over by the new owners. The photos were taken at Bickendorf airfield near Cologne (Köln).

Below: AEG G.IV closeup showing the landing lights in the lower nose and the numerous bombs mounted under the wings and fuselage. The finish is the typical AEG lozenge night camouflage.

Above: This AEG G.IV has lozenge night-bomber camouflage.

Below: This may be an early production AEG G.IV based on the light finish.

Above & Below: These front and rear views show how compact the AEG G.IV design was. The low frontal area to minimize drag was compromised by the multitude of struts and bracing wires.

Below: An AEG G.IV rests on an airfield with a Gotha GL.VII (left) at Bickendorf postwar.

Above: AEG G.IV with a ferocious face and heavy bomb load wears late-war insignia and "VII" on the fuselage side.

Below: AEG G.IV G1125/16 was given captured aircraft number G.105.

Above: AEG G.IV with shark mouth at the French aviation test center at Villacoublay shows its lozenge night-bomber camouflage typical of most G.IV bombers.

Below: AEG G.IV G.574/18 in Canada's national air museum is the only surviving example of the type and the only surviving multi-engine German aircraft of WWI.

Above: AEG G.IVb G.189/16 was one of the first production batch rebuilt with extended, three-bay wings to carry heavier bomb loads. The reason for the larger wing, a 1,000 kg bomb, is being loaded.

Below: AEG G.IV surrounded by British troops after capture.

Above: This front view shows how compact the AEG G.IV design was.

Above: AEG G.IV of *KG*4 in 1918.

Right & Below Right: The national insignia of these AEG G.IV bombers were cut out after capture.

Below: This AEG G.IV was repainted in French markings after capture. The Roman numeral 'III' may indicate it was assigned to *Bogohl* 3.

AEG G.IV G.1114/16 of *Kagohl* IV after capture on the Italian Front and seen during repainting in Italian colors; repainting with Italian markings was not yet complete.

AEG G.IV G.1125/16 was accepted for service on 10 Nov. 1917 and went to *Kagohl* III, *Kampfstaffel* 15. On the night of 23 Dec. 1917 it was downed by AA fire and landed intact at Achlet-le-Grand. It was flown to England for study where it was given capture number G.105.

AEG G.IV G.581/17, unit unknown, April–June 1918.

AEG G.V

In November 1917 AEG began work on a night bomber designed to carry a useful load of 2,100 kg, including a 1,000 kg P.u.W. bomb. The new bomber was based on the earlier AEG G.IVb-lang but with wingspan and area further increased for greater payload. The engines reverted to the 260 hp Mercedes D.IVa that powered production AEG G.IV bombers. The longer fuselage and box tail of the AEG G.IVb-lang were retained for improved engine-out handling.

To further improve flying qualities, the G.V used the Flettner *Hilfsruder* (servo rudder) that aerodynamically reduced the control forces and made large bombing aircraft less tiring to fly. Developed as part of Anton Flettner's work on automatic guided-missile control, the Flettner servo rudder was an advanced design feature of all late-1918 German bombers. According to one historian "the British Air Ministry technical reports of 1919 indicate complete ignorance of the aerodynamic principles employed!"

In May 1918 the AEG G.V prototype reached 4,000 meters in 70 minutes carrying a useful load of 2,100 kg. *Idflieg* held a competition between the AEG G.IV, AEG G.V, Friedrichshafen G.IV, and Gotha G.Vb bombers and found that, when carrying a useful load of 1,600 kg, the AEG G.V and Friedrichshafen G.IV were comparable in performance, and both were superior to the AEG G.IV and Gotha G.Vb in speed and climb. The first order for the G.V was placed in April 1918 and the first nine production G.V bombers were delivered in August. Between August and October 1918 a total

Above & Below: Similar views of AEG G.V 301/18 showing its enlarged, three-bay wing with Flettner tabs to reduce the pilot's aileron control forces. The 'box' tail with twin rudders gave better engine-out handling.

Above: AEG G.V anchors a display of late-war aircraft, perhaps postwar.

Above: The Fokker V17 monoplane fighter prototype provides an interesting contrast to the AEG G.V and emphasizes the size of the G.V.

Above & Below: On November 29, 1918, less than three weeks after the end of the war, AEG sold six AEG G.V bombers to Sweden. In fact, the first G.V, 1708/18, had already arrived at Eksjö on 23 November!

of 37 G.V bombers were delivered from the 50 that were ordered.

Postwar, AEG sold six AEG G.V bombers to Sweden on November 29, 1918, with AEG to deliver the aircraft by air. The first G.V, 1708/18, had already arrived at Eksjö on 23 November 1918. Three G.V bombers left Hennigsdorf on 29 March 1919 but two force landed before reaching Sweden and the third was destroyed on take-off in Vilemölla on 4 April 1919. In May–June 1919 two additional G.V bombers (including G.1712/18) were flown to Kristianstad, followed by two further replacements (including G.1710/18) in August and September 1919.

Above & Below: Additional views of AEG G.V 644/18 showing more detail of its ailerons with Flettner tabs to reduce the pilot's aileron control forces. The AEG G.V was the best two-engine German bomber in service at war's end. For some reason the port engine of this aircraft has been removed.

Above: Side view of an AEG G.V of the second batch showing the Flettner tabs and 'box' tail with twin rudders.

Below: Rear view of an AEG G.V showing the 'box' tail with twin rudders to better advantage.

Above: Front view of AEG G.V 644/18 emphasizing the three-bay wings and distinctive Flettner tabs on the ailerons. The port engine has been removed, perhaps for maintenance.

Below: Side view of an AEG G.V 644/18; power was from a pair of 260 hp Mercedes D.IVb engines like the earlier G.IV.

Facing Page: German bombers photographed postwar. The two aircraft at the top of the photo are AEG G.V bombers. A Gotha G.V is at bottom left and another AEG G.V is at bottom right.

Above: AEG G.V undergoing adjustment of its ailerons and Flettner tabs.

Above: AEG G.V after its crash in Sweden in 1919 displays under-fuselage details not normally visible in photos. Swedish markings have already been applied over the German night-bomber printed camouflage fabric.

Swedish Adventure by Colin Owers

Five AEG G.V bombers were purchased by Sweden and received the Swedish Army serials: 8501 (that became 8502 as the army used odd numbers for single-seaters the original allocation was in error), 9504, 9506, 9508, 9510 and 9512. All were delivered and flown occasionally in 1919 by the Army Aviation Company as reconnaissance aircraft in their original lozenge fabric with Swedish markings painted over the German markings. They were delivered without any armament or radio equipment.

G.V 1708/18 was the first aircraft delivered. It flew non-stop from Berlin to Eksjö in Southern Sweden, on 23 November 1918, on a "test flight". The Swedish Army was interested and purchased the machine and ordered five more. DLR was the intermediary and the order was placed with this concern. G.V 1708/28 was given the Swedish Army serial 8501 but this was replaced by 8502 in March 1919. This machine made 24 flights but only one long distance one before crashing on 14 May 1919.

The first three of the order left Berlin for Sweden on 19 March.

The next machine's original serial is unknown. It suffered at least two forced landings on the delivery flight. It was 9504 in Swedish service. This was the only G.V to crash after a five minute flight, its first in Swedish service, on 19 May 1919, the crew escaping without injury. The aircraft ended up on its nose and was the subject of many photographs.

The third machine is also unknown. It crashed during landing at Warnemünde Harbour on 28 March 1919, killing the crew. The wreck was salvaged that evening.

The fourth machine also suffered force landings. Oil failure caused a force landing at Vitemöllä on 30th. When attempting to continue their journey the machine crashed on 4 April. The German crew were unhurt. AEG sold the remains to the Swedish Army as junk. The Swedes were interested in the motors that had come through the accident without damage. It is probable that this aircraft was repaired and became 9512. Aircraft 9512 was written off on 1 June 1920.

AEG G.V No.5 was 1712/18 and left on 18 May 1919, and after two force landings arrived on 21st. As Army 9506 it had no recorded flights at Malmslätt, and was written off on 16 December 1919, and finally scrapped in 1920.

AEG G.V No.6 is unknown. It arrived on 6 June 1919, suffering only one force landing on its delivery trip. Again, as 9508, no flights were recorded for this machine at Malmslätt. As 9508 it was written off on 3 April 1922.

G.V 1710/18 was the next AEG delivered. It became 9510. It had suffered a forced landing and crash on its delivery flight due to a broken motor shaft. It arrived on 10 August 1919, and was handed over the following day. It crashed on 13 August 1919, three days after arrival while still wearing its German markings. It never wore its Swedish serial 9510. It was officially written off on 13 December.

The last AEG G.V (erroneously recorded as 501/18) apparently had no problems on its delivery flight but was not accepted by the Swedish control officer due to its shoddy workmanship. It was destroyed on the ground on 24 November 1919 in a storm at Kristianstad.

AEG G.Vs 8592 (sic), 9504, and 9512 were written off on 13 December 1919, in order to clear up the paper work. The AEG bombers may have been too advanced for the Swedish pilots used to single-engined machines; however, it appears that there were problems with quality control at the AEG end. The 260 hp Mercedes engines proved useful and were used in the locally produced FVM S 18 reconnaissance aircraft.

Note: This section has benefited from the research of Lennart Andersson. Lennart and Ray Sanger have completed a book on the fate of the German aircraft post-WWI entitled *Retribution and Recovery. German Aircraft and Aviation 1919–1923.*

Below: AEG G.V 9504 after its crash in Sweden in 1919.

AEG G.V G.625/18

AEG G.V G.644/18

AEG G.V 9504 in Swedish service postwar

Above & Below: AEG G.V in Sweden postwar. The snow indicates this may be the first aircraft 1708/18, delivered 23 November 1918 before the purchase contract was signed.

Above: AEG G.V during its delivery flight to Sweden.

Left: Crash of AEG G.V 1710/18 in Sweden Aug. 13, 1919.

Above: AEG G.V 301/18 in American hands postwar.

Below: Rear view of an AEG G.V highlighting its massive wing span and 'box' tail.

AEG N.I

Above: The AEG N.I was a long-span version of the AEG C.IV modified to carry more bombs, and this view clearly shows the key N.I recognition features. First, the N.I had a long wing with 3-bay bracing compared to the shorter, 2-bay wings of the C.IV. Next, the triangular bracing truss above the upper wing center section was required to strengthen the longer N.I wing against bending so it could pass the required load test. Unlike its daylight predecessor the AEG C.IV, the N.I did not have a fixed machine gun for the pilot; the observer's gun was thought sufficient for both strafing and defending the aircraft at night. The high exhaust stack directed the noise and flames away from the aircraft, which was important to maintain the pilot's night vision. Typical AEG sprayed camouflage on the upper surfaces is evident.

Like the General Electric Company in the United States, the Allgemeine Elektrizitäts-Gesellschaft was one of the more powerful and progressive technology-based companies in Germany. As part of its forward thinking, AEG established a *Flugtechnische Abteilung* (aero-technical department) at Hennigsdorf (north of Berlin) in 1910 directed by *Oberingenieur* Paul Stumpf, who pioneered the use of autogeneous-welded steel tubing for aircraft, an advanced structural technology at a time when most airplanes were made of wood. All subsequent AEG aircraft used this technology.

New assembly shops were opened in July 1915 in Nieder-Neuendorf to become the AEG Flugzeugfabrik led by *Direktor* Bassler. In the winter of 1915–1916, flight tests of an improved two-seat reconnaissance biplane, known as the AEG C.IV, were satisfactorily completed. Ordered in quantity, the C.IV was praised as a fast, rugged aircraft that stood up well during combat operations.

AEG responded to the N-type specification by modifying their successful AEG C.IV to lift a 300 kg bomb load by increasing the wingspan 2.24 meters (7.3 ft); changing the wing from a two-bay to a three-bay design. In September 1916, the AEG C.IVn prototype completed its initial flight trials, proving that it was stable and easy to fly, important criteria for night flying. AEG received a production order in December 1916 for 100 AEG C.IVn night bombers.

AEG N.I Specifications

Engine:	150 hp Benz Bz.III	
Wing:	Span Upper	15.24 m
	Span Lower	14.62 m
	Chord Upper	1.65 m
	Chord Lower	1.65 m
	Gap	1.95 m
	Area	41.38 m^2
General:	Loaded Weight	1,609 kg
Maximum Speed:		143 kmh
Climb:	1000m	10 min
	2000m	23 min
	3000m	50 min

The first AEG C.IVn production example (C.9323/16) was dispatched to Adlershof in April 1917 for type-testing. The C.IVn wing failed repeated load tests and it was not until 7 June 1917 that sufficient bending strength was achieved by the installation of triangular reinforcing trusses over both the forward and aft center section spars. This solution, structurally efficient but aerodynamically inelegant, sufficed for a slow aircraft not expected to engage in air-to-air combat.

Timing of the initial deliveries of the AEG C.IVn and when it first reached the front are not known

Above: AEG C.IV C.1042/16 with typical AEG sprayed camouflage is probably the prototype C.IVn because C.1042/16 is a normal C.IV number. This aircraft appears to lack the characteristic over-wing bracing used on C.IVn/N.I production aircraft and retains the 160 hp Mercedes D.III from the C.IV. Three 50 kg P.u.W. bombs are visible under the port lower wing.

because it was counted among the regular AEG C.IV aircraft in the *Frontbestand*. However, by August–September 1917 the *Nachtflugzeug* (N = night aircraft) category had been established by *Idflieg,* and the first two AEG N.I bombers were recorded at the front in October 1917. *Idflieg* approved the second production order for 100 AEG N.I bombers in November 1917.

A reliable, efficient aircraft, the AEG N.I was primarily assigned in ones or twos to divisional two-seater and bombing units to perform short-range raids behind the enemy lines. Later in the war, some N.I biplanes served as advanced trainers.

B.II AEG 191 from the AEG Flugplatz Nieder-Neuendorf

Above & Facing Page: These views show the N.I's characteristic triangular bracing trusses above the upper wing center section. The trusses strengthened both wing spars to handle the increased bending forces of the longer wing. Multicolor, hexagonal camouflage similar to that used by the AEG twin-engine night bombers was applied to many N.I aircraft. Given that the N.I shared the same operational role, it is not surprising that similar camouflage was used. At right, "Franz" and "Emil" look less than enthused about their next night bombing mission. A Wolff propeller is fitted to this N.I.

Below: This rear view shows the two key AEG N.I recognition features; the long, 3-bay wing and the triangular braces for both front and rear spars above the center section. The hexagonal night camouflage typical for these aircraft is evident.

WOLFF

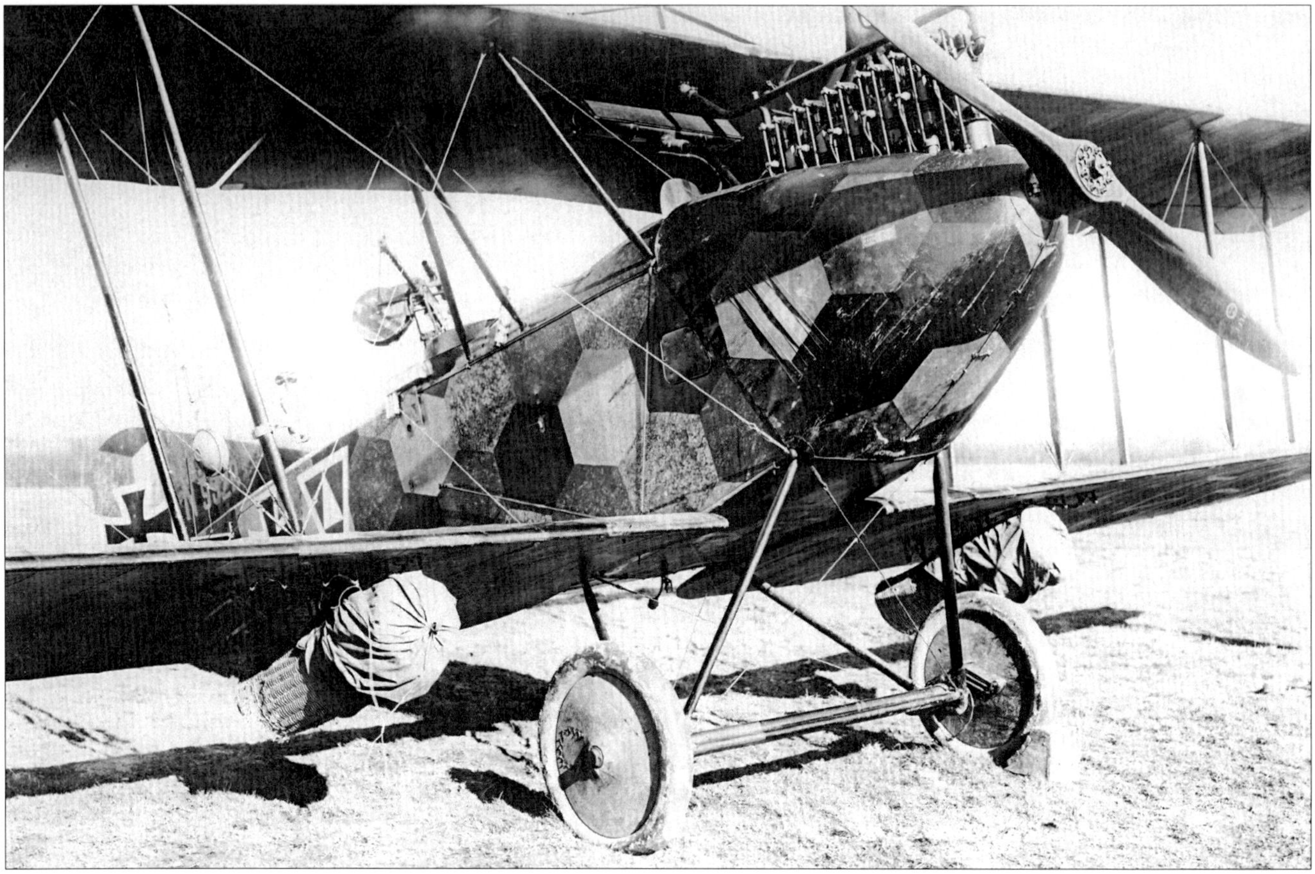

Above: AEG N.I N.9323/16 carries air-dropable supply containers under its wings; the parachutes are wrapped around the front of the wicker containers. One of the few N.I bombers seen in unit markings, it is assigned to *FA(A)* 209. The night bomber hexagonal camouflage shows the wear from operational flying. A Wolff propeller is fitted, and the landing light in the leading edge of the upper left wing is visible.

Below: Two AEG N.I bombers are seen in in this photo of *FA(A)* 209. The aircraft on the far right in the row nearest the camera is a Rumpler C.IV; to its left in the row nearest the camera are two AEG N.I night bombers.

Above: AEG N.I N.9389/16 (possibly N.9399/16; the next to last digit is partly obscured) wears the two-color sprayed camouflage scheme and has lasted long enough for Balkan crosses to replace the original iron cross insignia. Wooden wheels are fitted due to the scarcity of rubber, and the landing light in the leading edge of the upper left wing is visible. The wood wheels indicate this machine may have been in training service when the photo was taken.

Below: Two AEG N.I bombers wearing the hexagonal camouflage scheme have collided on the ground. N.9417/16 is nearest the camera and has the letters 'AEG' in a rectangle on the lower rudder.

Above: AEG N.I 139/17, the only known photograph of an N.I from the second production series, shows the extended cabane section and the cowling panels that fold back to expose the engine for maintenance. This photograph was taken at the Bavarian training base in Lechfeld, and the aircraft is fitted with a 180 hp Argus As.III engine. National insignia is in the late style and the camouflage was sprayed on in two colors.

Below: Some AEG N.Is were used briefly in a civilian capacity after the war. Civil N.I aircraft *may* have been given the 200 hp Benz Bz.IV engine, which could account for the different shape of the engine cowling and over-wing radiator. The wing structure appears to be modified.

Above: A unit portrait was taken in front of this AEG N.I; unfortunately, no further details are known. The landing lights in the leading edge of the upper wings are clearly visible.

Below: Crash of an unknown AEG C.IVn/N.I showing the night camouflage and upper-wing bracing. The night bomber camouflage is so dark it is hard to see the national insignia. The two trusses bracing the upper wing spars are prominent.

AEG N.I in Color

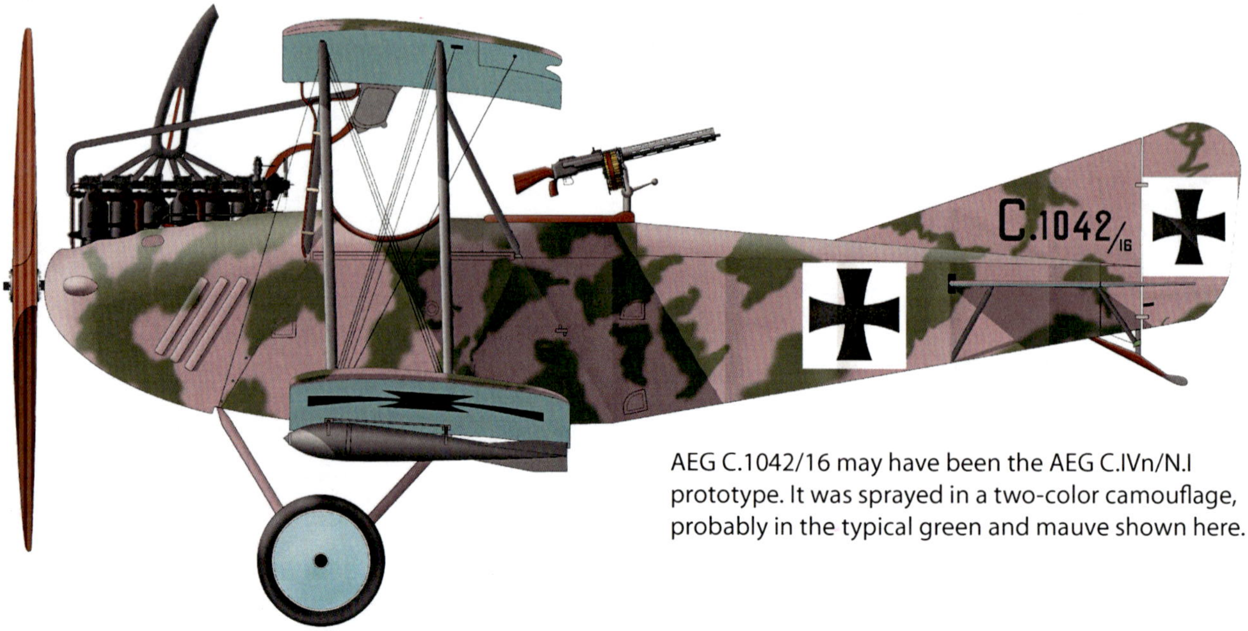

AEG C.1042/16 may have been the AEG C.IVn/N.I prototype. It was sprayed in a two-color camouflage, probably in the typical green and mauve shown here.

AEG N.I N.9417/16 is typical of many for which photographs are available. National insignia with little or no white borders are applied over the typical overall AEG dark, hexagonal night camouflage as seen on twin-engine AEG night bombers. The serial number is present in light paint and the AEG label is on the rudder below the cross.

AEG N.I N.9323/16 is virtually the only N.I for which we have a photograph showing both the serial number and a unit insignia. This aircraft was assigned to *FA(A)* 209.

AEG N.I N.9389/16. The reference photo shows a sprayed finish that appears to be in two colors. The two colors used here were based on the colors used by the AEG J.I and J.II that were contemporary with the N.I.

Plan view of AEG N.I in typical AEG hexagonal night bomber finish.

AEG N.I N.139/17. The reference photo shows a sprayed finish, but whether of two or three colors is uncertain. The two colors used here were based on the colors used by the contemporary AEG J.I and J.II and the pattern is representational.

AEG R.I & R.II

In late 1916 AEG received an order for two giant bombers, R.I 21/16 and 22/16, and established an R-plane department on Jan. 1, 1917. The R.I wing spars were made of steel and the ribs were duralumin instead of wood like smaller AEG designs. The fuselage construction was fabric-covered steel tubes like other AEG aircraft. Five machine gun positions were provided.

AEG R.I Specifications		
Engines:	4x260 hp Mercedes D.IVa	
Wing:	Span	36.0 m
	Chord	3.8 m
	Gap	4–4.5 m
	Area	260.0 m^2
General:	Empty Weight	9,000 kg
	Loaded Weight	12,700 kg

Like some other R-plane designs, AEG used a centralized power system design with engines in the fuselage driving two propellers via shafts. The choice of centralized power was driven by the inability to feather the propellers of the time in case of engine failure, the resulting drag of the wind-milling propeller severely limiting the ability of the aircraft to maintain altitude after engine failure. With a centralized design, a failed engine could be de-coupled from the propeller by a clutch, allowing the running engines to power all the propellers. The R.I was powered by four 260 hp Mercedes D.IVa engines.

The R.I first flew on June 14, 1918 after successful engine ground tests on May 29 and 30. Piloted by *Oblt*. Brückmann, the R.I flew for 27 minutes. The R.I was very tail heavy; even full down elevator was insufficient to maintain level flight and landing was made only by shifting all crewmembers to the nose. In addition to center of gravity problems, flight tests showed that rudder area and cooling were inadequate. Modifications were made quickly and the aircraft was returned to flight test. Brückmann flew the modified R.I on Sept. 3, 1918 despite warnings that the glue used in the new AEG propellers had not had time to set. After about a hour in the air, one propeller disintegrated, causing its transmission shaft to thrash around and damage the center section structure. The R.I disintegrated in flight killing all seven crewmen aboard. The fatalities included Brückmann, *Lt*. Otto Reichardt, and *Lt. Dr.* Oskar Wittenstein, all valuable members of the *Kommando Riesenflugzeug Abteilungen. Idflieg* financed assembly of R.I 22/16 through January 1919 to make work for AEG personnel but the bomber was never completed. Some components for the R.I production batch of six bombers, R.59 to R.64, were partially completed at the Armistice.

In September 1917 AEG submitted a proposal to *Idflieg* for a giant monoplane powered by four

Below: The massive AEG R.I 23/16 in its final form. (The Peter M. Bowers Collection/The Museum of Flight)

Above & Below: The AEG R.I 23/16 as completed with four-blade propellers. (The Peter M. Bowers Collection/The Museum of Flight)
Bottom: The AEG R.I 23/16 in final form with two-blade propellers and new radiators on the lower center-section struts. (The Peter M. Bowers Collection/The Museum of Flight)

500 hp engines, manned by a crew of ten, and mounting six machine guns. This project was approved and assigned the designation AEG R.II 205/17. It is probably the same AEG monoplane bomber with a wingspan of 45.9 meters for which Daimler was to supply the engines and drive assembly. Only preliminary design work had been completed by the Armistice.

Miscellaneous AEG Aircraft

AEG-Wagner Monoplane

In late 1914 the AEG-Wagner monoplane was built by the AEG Flugtechnische Abteilung in Hennigsdorf to evaluate the wing shape designed by *Ingenieur* Wagner. The purpose of the design may have had something to do with using the propeller wash to develop additional lift. The first model and five other aircraft were destroyed in a hangar fire in September 1914, although the steel-tube fuselage and rotary engine survived. A second AEG-Wagner monoplane appeared in early 1915. Flight trials demonstrated little advantage of the Wagner wing and the second AEG-Wagner monoplane was hung from the rafters in the AEG factory for storage. The fuselage and vertical tail surfaces are presently in the Polish aircraft collection at Krakow.

AEG Igel Project

On July 22, 1916, *Idflieg* issued specifications for a two- or three-seat reconnaissance machine powered by a 500 hp Mercedes engine (called the Igel-hedgehog) buried in the fuselage. A climb to 4000 meters in 24 to 26 minutes and a top speed of 215 km/h were specified. The buried Igel engine would drive two outrigger-mounted propellers at 900 rpm. A nose and rear gun position provided defensive armament. To speed development it was proposed to build different wing cellules for comparative trials. *Idflieg* preferred a triplane configuration to achieve a high climb rate. AEG was chosen to design and build a prototype because "their recent aircraft have been good and steel tube construction was especially amenable to experimentation." Albatros was not considered because they were already fully busy with new designs, and *Idflieg* judged the task to be too difficult for other firms. The drawings for the Igel project biplane were completed by AEG in October 1916. It is unknown if assembly work was started, but it is certain the type was never completed.

AEG Naval R-Planes

On 12 October 1917, the German Navy ordered a giant AEG R-flying boat (Marine Number 9301, formerly Marine Number 2139) powered by four 245 hp Maybach Mb.IVa engines. The AEG flying boat project was listed as 'in design' until the project was canceled on November 26, 1918.

Two AEG R-floatplanes (Marine Numbers 9302 and 9303), powered by four 245 hp Maybach Mb.IVa engines, were ordered by the German Navy on January 23, 1918. The giant seaplane was basically an AEG R.I fitted with floats. The AEG 9302 fuselage, control surfaces and metal fixtures were under construction in June. At the time of cancellation on December 18, 1918, the AEG 9302 fuselage was in final assembly and the floats and beaching trolley, both manufactured by the Zeppelinwerk Staaken, were completed. Only the fuselage parts and metal fittings for AEG 9303 were finished at the time.

Above Right: AEG B.II showing its 150 hp Benz Bz.III engine.

Right: AEG B-type trainer '43' after a bad landing.

AEG Aircraft in Retrospect

One of the largest and most capable engineering companies in Germany, AEG became one of the most important aviation manufacturers during WWI. AEG's trademark was robust, welded steel-tube structures that gave its aircraft the strength and resilience to withstand the rigors of combat flying in a demanding environment. Among German manufacturers only AEG and Fokker used this structural technology.

In contrast to its innovative structural methods, AEG's aerodynamic approach was conservative. AEG wartime aircraft were nearly all conventional biplane designs with a few experimental triplanes built during Germany's 'triplane craze'. However, all production AEG warplanes were biplanes.

AEG's reconnaissance two-seater designs were moderately successful. Only one, the AEG C.IV, was produced in significant numbers and was the most numerous AEG type. The C.IV design was of more significance than that because it also formed the basis for the AEG N.I single-engine night bomber and the J.I and J.II armored two-seat infantry cooperation and ground-attack aircraft.

If AEG found only moderate success building B-type and C-type aircraft, it was more than successful in two other important areas, two-engine bombers and armored infantry aircraft.

AEG is best known for its two-engine bombers, the most numerous of which was the G.IV. Not as famous as the Gothas that gained notoriety for bombing London, the AEG G-types were robust aircraft that were much easier to fly than the Gothas. AEG's initial design for the G-types was well-conceived and the basic configuration was developed throughout the war by mounting more powerful engines, strengthening the structure, and enlarging the wing for greater payload. By the time AEG's last and largest bomber, the G.V, arrived, development of Flettner servo tabs greatly reduced aileron control forces, making it much easier for the pilot to handle despite its size. In addition, the G.V also introduced the 'box' tail that improved engine-out controllability. The combination of these innovations plus the large bombload made possible by the extended wing made the AEG G.V probably the best two-engine German bomber of the war.

On the other hand, AEG was late getting in the R-plane business and only one AEG R-plane flew, which was destroyed in an unfortunate accident. The fact it was so susceptible to propeller failure was a shortcoming of its centralized power design.

In addition to AEG's success in two-engine bombers, the AEG J.I and J.II were particularly successful and operationally important aircraft. The J.I was the first German armored warplane to reach the front, and the J.II was built in greater numbers than any other German armored warplane. The basic design, developed from the C.IV, lent itself to continuous development. While the Junkers J.I is famous for being the best German armored infantry cooperation warplane and for being the world's first all-metal production aircraft, it was too heavy to accommodate the additional armament found important for ground-attack. In contrast, the AEG J.II was built in greater numbers because it was easier to build and also because it was able to carry the additional weapons increasingly found necessary. Furthermore, the steel structure and doubled control cables and bracing wires of the AEG J.II made it nearly as resistant to ground fire as the Junkers. The J.II was the second most numerous warplane AEG produced but has not received the recognition it earned.

AEG combined its bomber and armored infantry plane experience to create the AEG G.IVk, WWI's equivalent of the modern anti-tank attack helicopter. Armed with small bombs and two flexible 20mm Becker cannon for anti-tank and general ground-attack use plus two flexible machine guns for protection against fighter attack, the AEG G.IVk was unique in WWI aviation.

One important area where AEG had mixed results was fighters. The AEG D.I fighter, which flew during mid-1917, was faster than any German fighter that reached the front during the war, a significant achievement. And its triplane derivative was the fastest German triplane fighter. However, two unexplained fatal crashes of prototypes caused *Idflieg* to cancel the D.I program. Certainly there were grounds for great concern and perhaps cancellation was the best decision. However, some D.I aircraft continued to fly, apparently without problems, bringing into question whether D.I development and production should have continued. Could the D.I have been both safe and a success had it been produced in quantity and reached the front? There is no question that German fighter pilots in late 1917 were in desperate need of a fighter of its performance. Unfortunately, the D.I never had an opportunity to prove itself in combat and we will never know whether or not it was safe enough to have succeeded; certainly it had the performance.

Above: This photo displays AEG G.II G.6/15. The late A.E. Ferko claimed that this was the plane in which Richthofen and Georg Zeumer often flew together.

Bibliography

Books

Gray, Peter, and Thetford, Owen, *German Aircraft of the First World War*, second revised edition, New York: Doubleday & Company, Inc., 1970.

Grosz, Peter M., *Junkers J.I*, Berkhamsted, Albatros Publications, 1993.

Lamberton, W.M., *Reconnaissance & Bomber Aircraft of the 1914–1918 War*, Letchworth, England, Harleyford Publications, 1962.

Articles

Grosz, Peter M., "An Armored AEG" *Windsock* Vol.11 Nr.6, Nov/Dec 1995.

Grosz, Peter M., "The 2 CM Becker Aircraft Cannon – Development and Use" *Over the Front* Vol.7 No.2, Summer 1992.

Grosz, Peter M., "Rare Birds: AEG J.Ia & J.II" *Over the Front* Vol.17 No.4, Winter 2002.

Herris, Jack, *German Armored Warplanes of WWI*, Aeronaut Books, 2012.

Herris, Jack, *Nachtflugzeug! German N-Types of WWI*, Aeronaut Books, 2012.

Rabe, Hans-Gerd, "Over the Lines in an AEG J.I," *Cross & Cockade* Vol.10 No.1, Spring 1969.

Internet

http://en.wikipedia.org/wiki/Duralumin

http://en.wikipedia.org/wiki/Deutsche_Luft-Reederei

Unpublished manuscript by Peter M. Grosz

Selected Index

August, Duke Ernst 97
Bassler, *Director* 150
Berthold, Rudolph 96, 97
Brückmann, *Oblt.* 160
Bülow, Walter von 106
Diemer, Fritz 131
Firle, Otto 79
Flettner, Anton 137
Hendrichs, Julius 47
Höhndorf, Walter 49
Johannes, Peter 58
Kantiz, Willy 92
Kieser, *Hptm.* 31
Köhl, Hermann 186
König, Georg 49
Milenz, *Lt.* 42
Rabe, Hans-Gerd 56, 58
Raetsch, *Vzfw.* 34
Reichardt, Otto 160
Richthofen, Manfred von 105, 164
Schauenburg, Theodor 52
Schulz, *Offz.* 34
Schwandt, Paul 122
Siegert, Wilhelm 83
Spiegel, *Lt.* 42
Stumpf, Paul 150
Wagner, *Ing.* 162
Wilhelm, Prince August 97
Wittenstein, Oskar 160
Zeumer, Georg 164

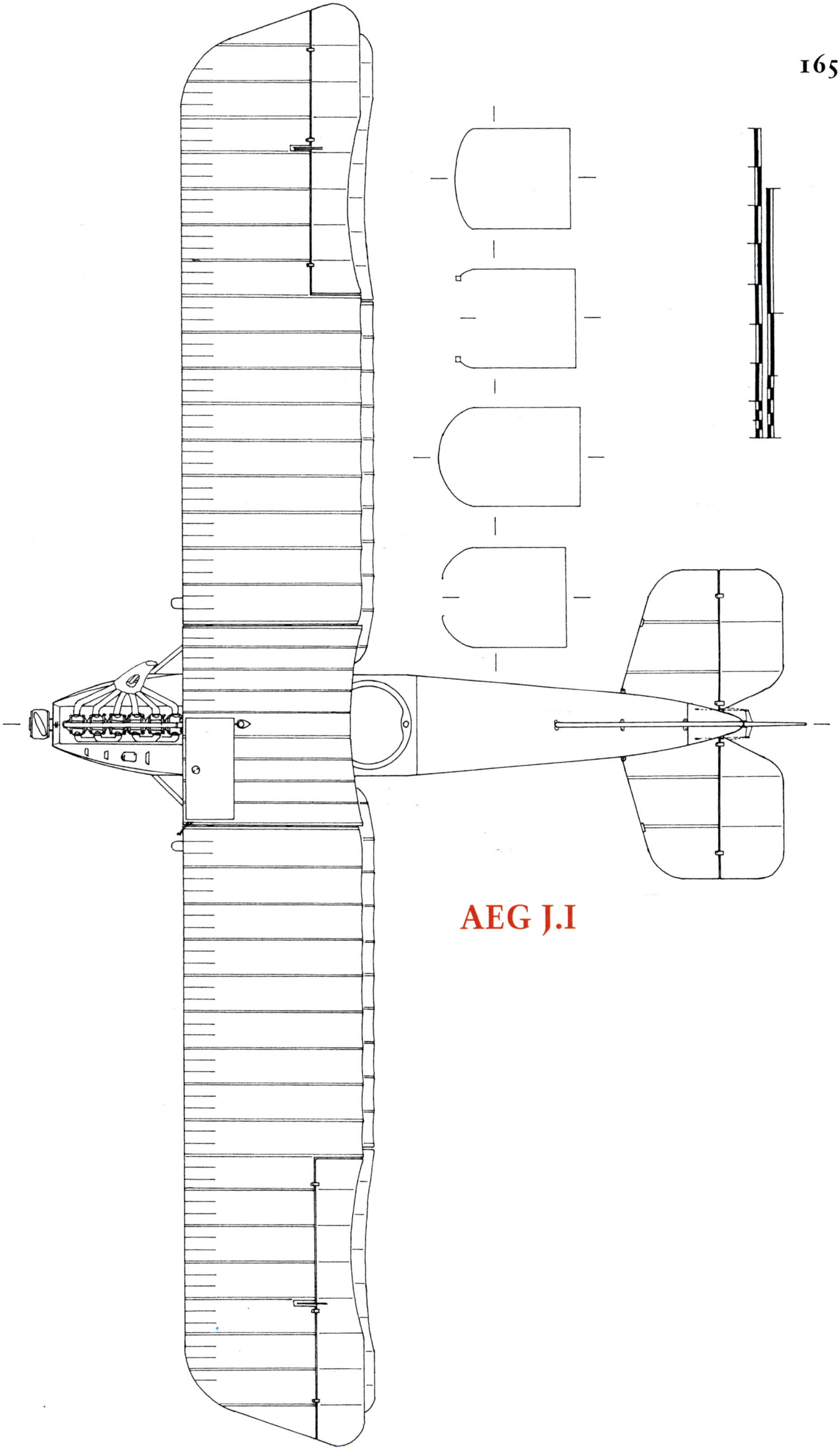

AEG J.I

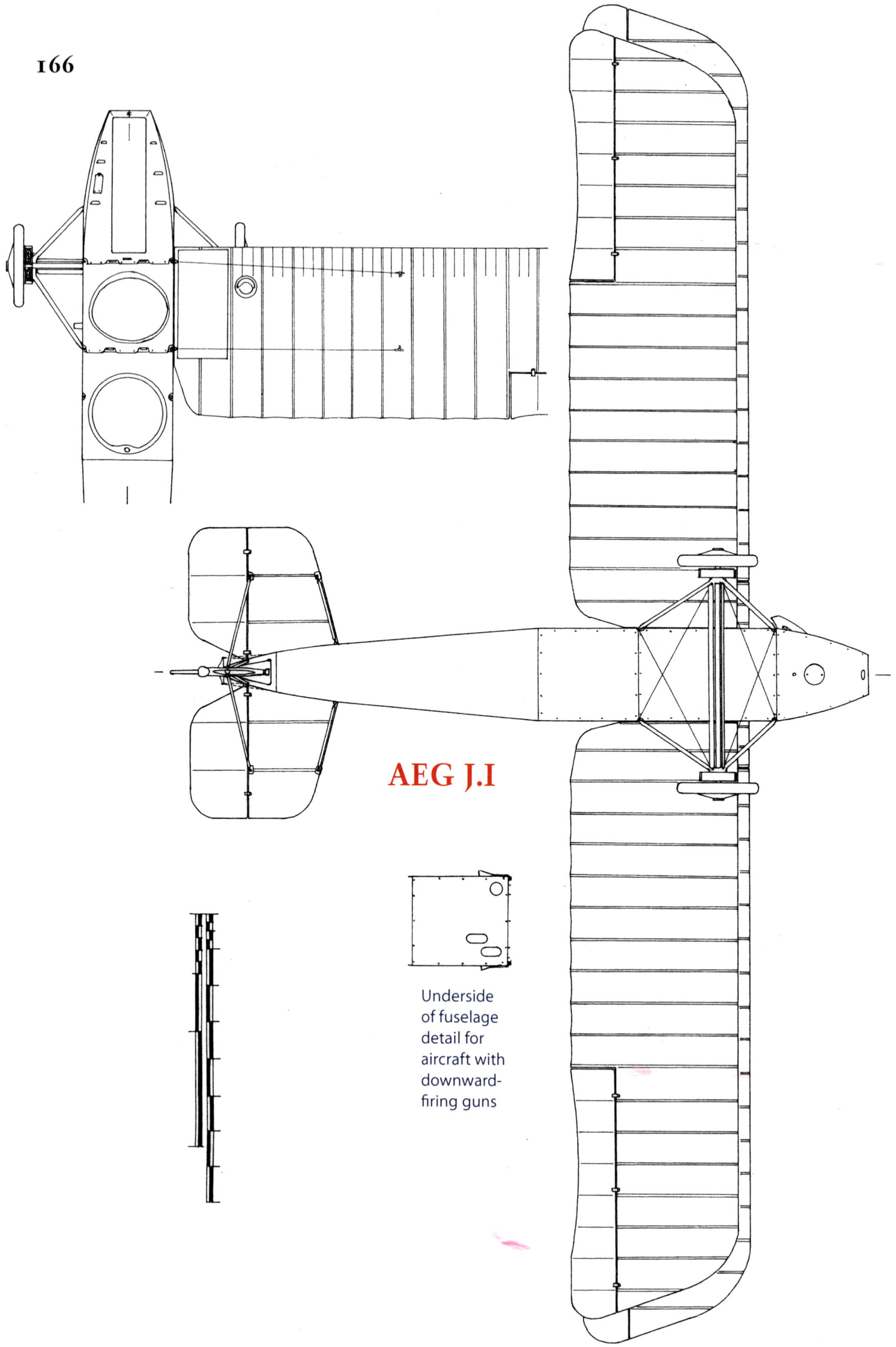

AEG J.I

Underside of fuselage detail for aircraft with downward-firing guns

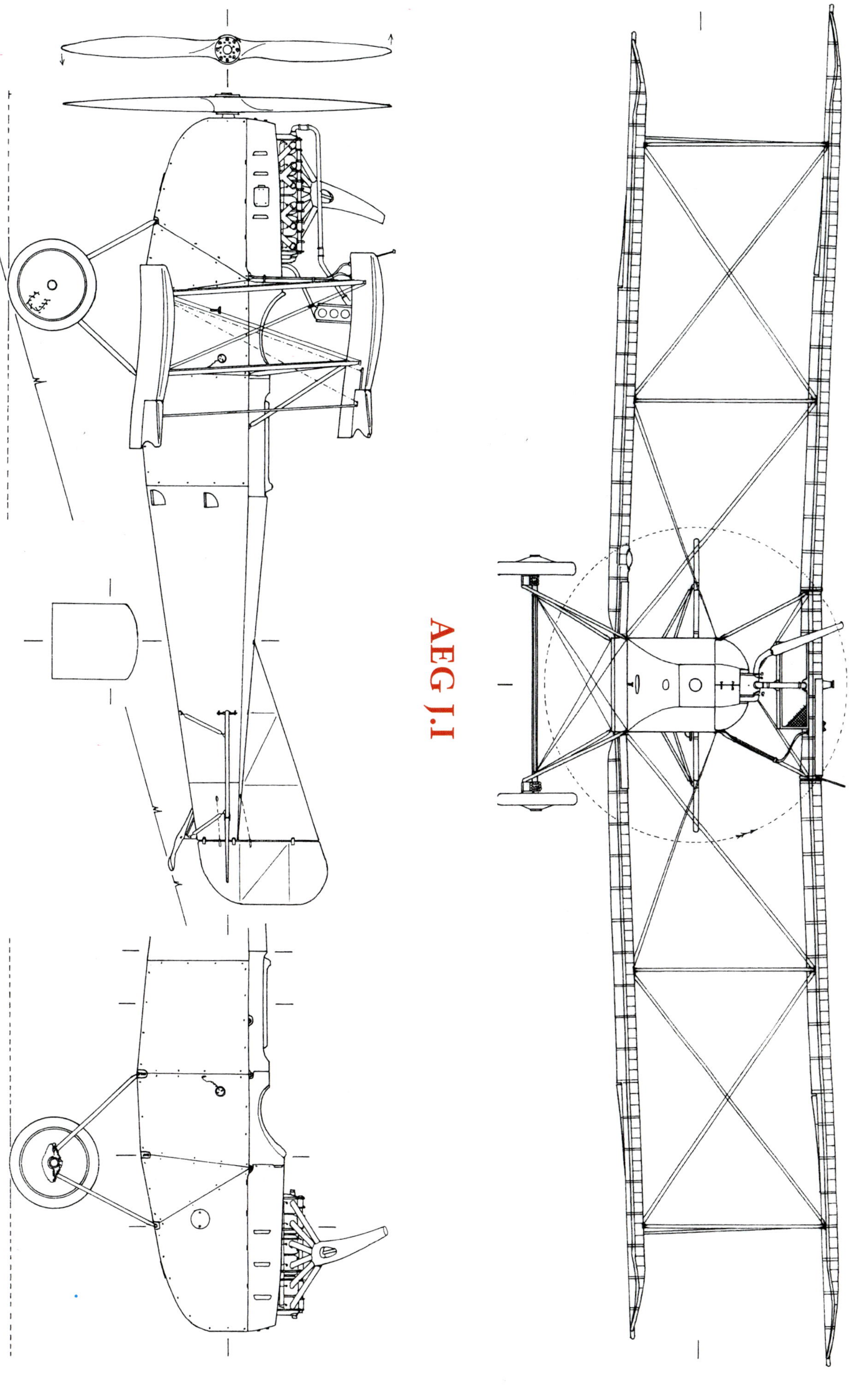

AEG J.I

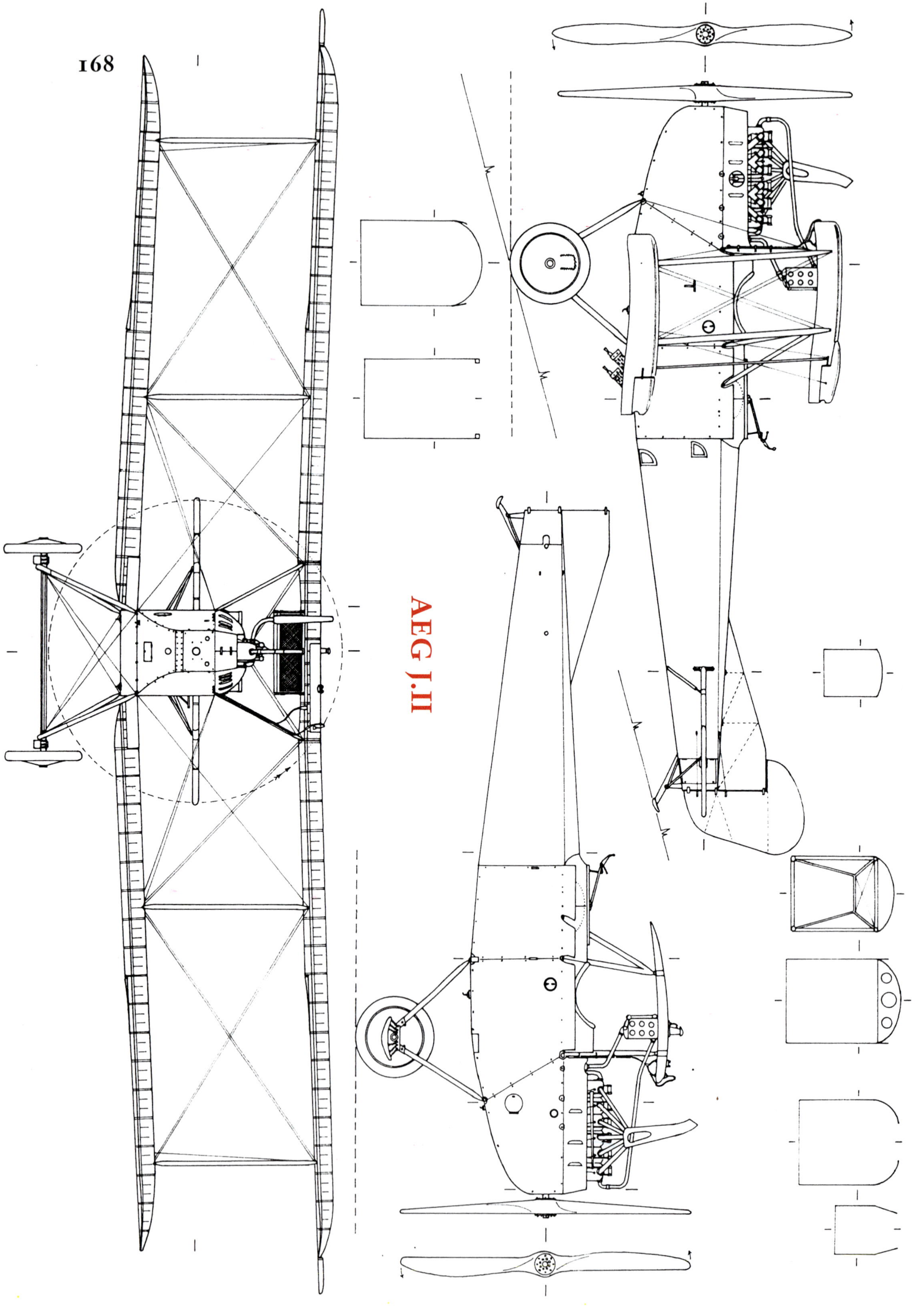

AEG J.II

AEG J.II

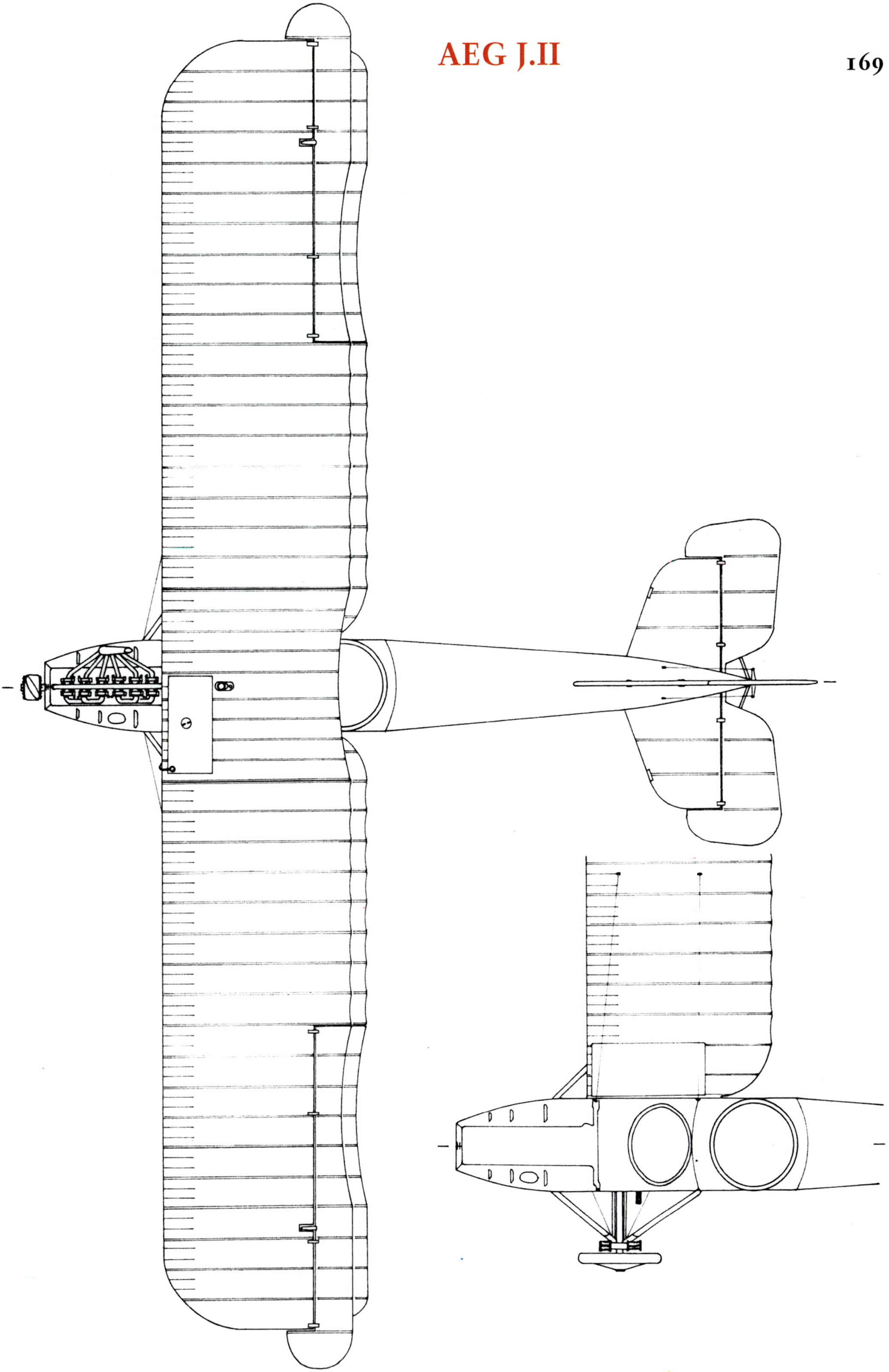

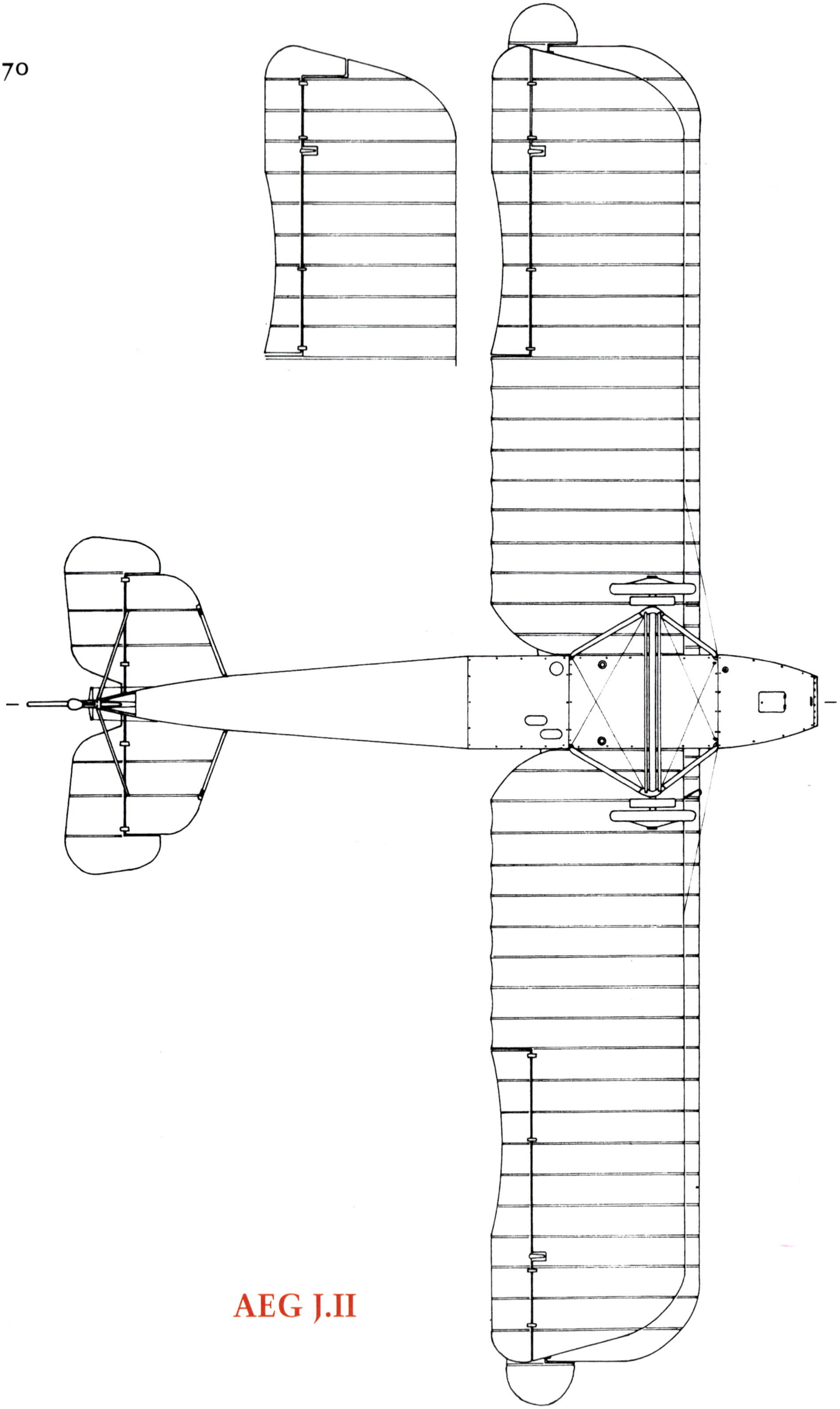

AEG J.II

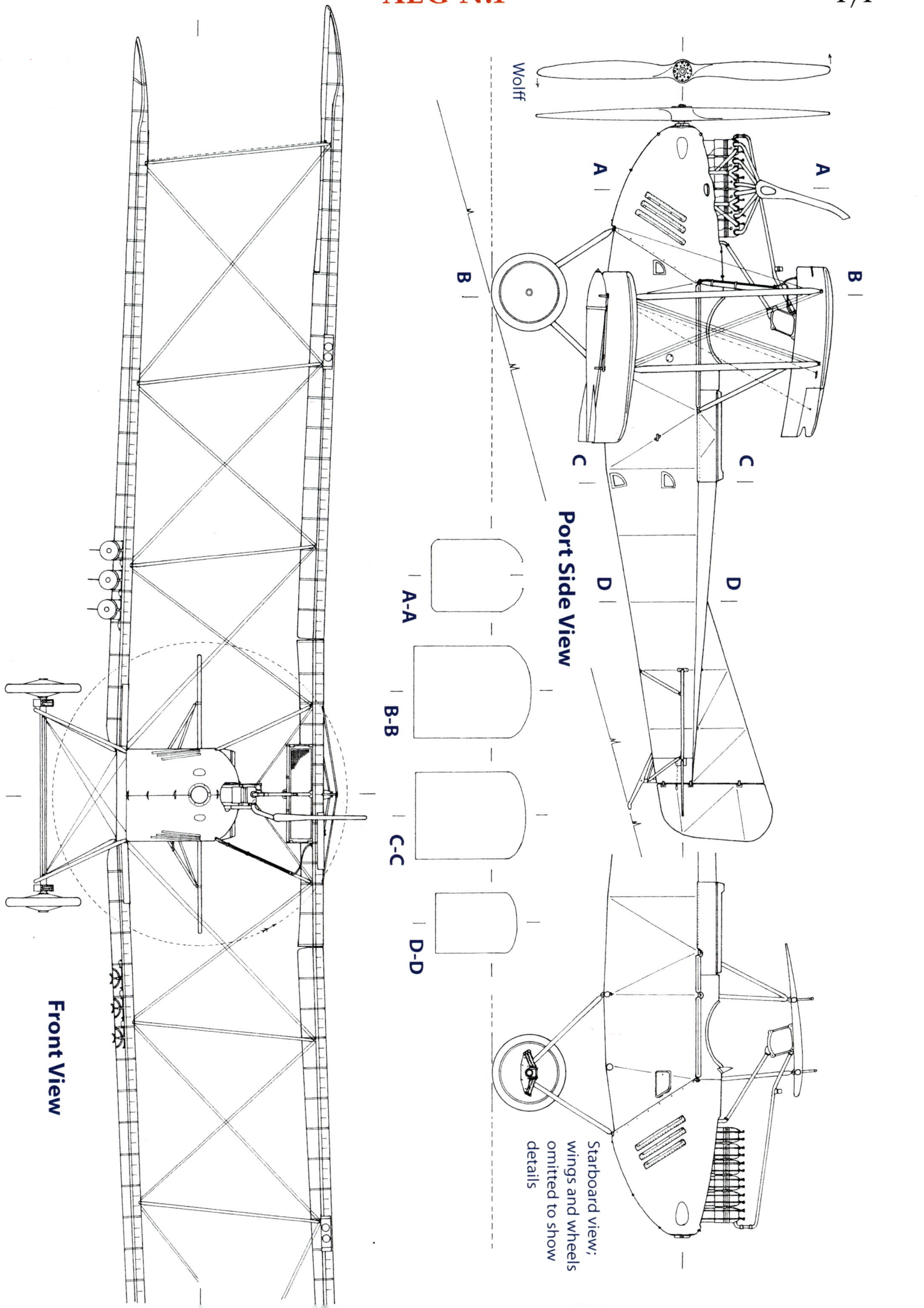
Wolff
A
A
B
B
C
C
D
D
Port Side View
A-A
B-B
C-C
D-D
Front View
Starboard view; wings and wheels omitted to show details

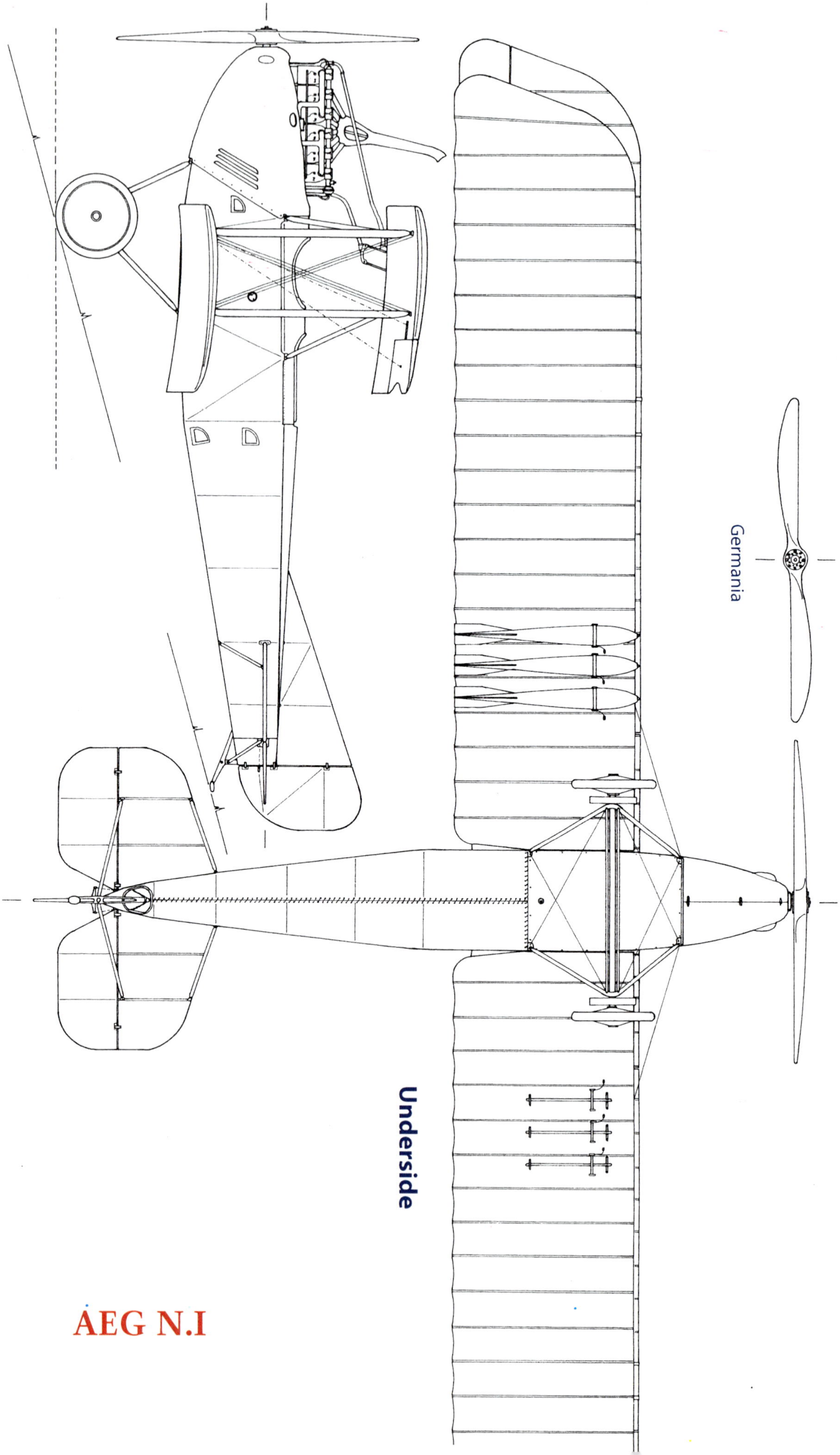

AEG C.IVn prototype
160 hp Mercedes D.III engine

AEG N.I

AEG N.I

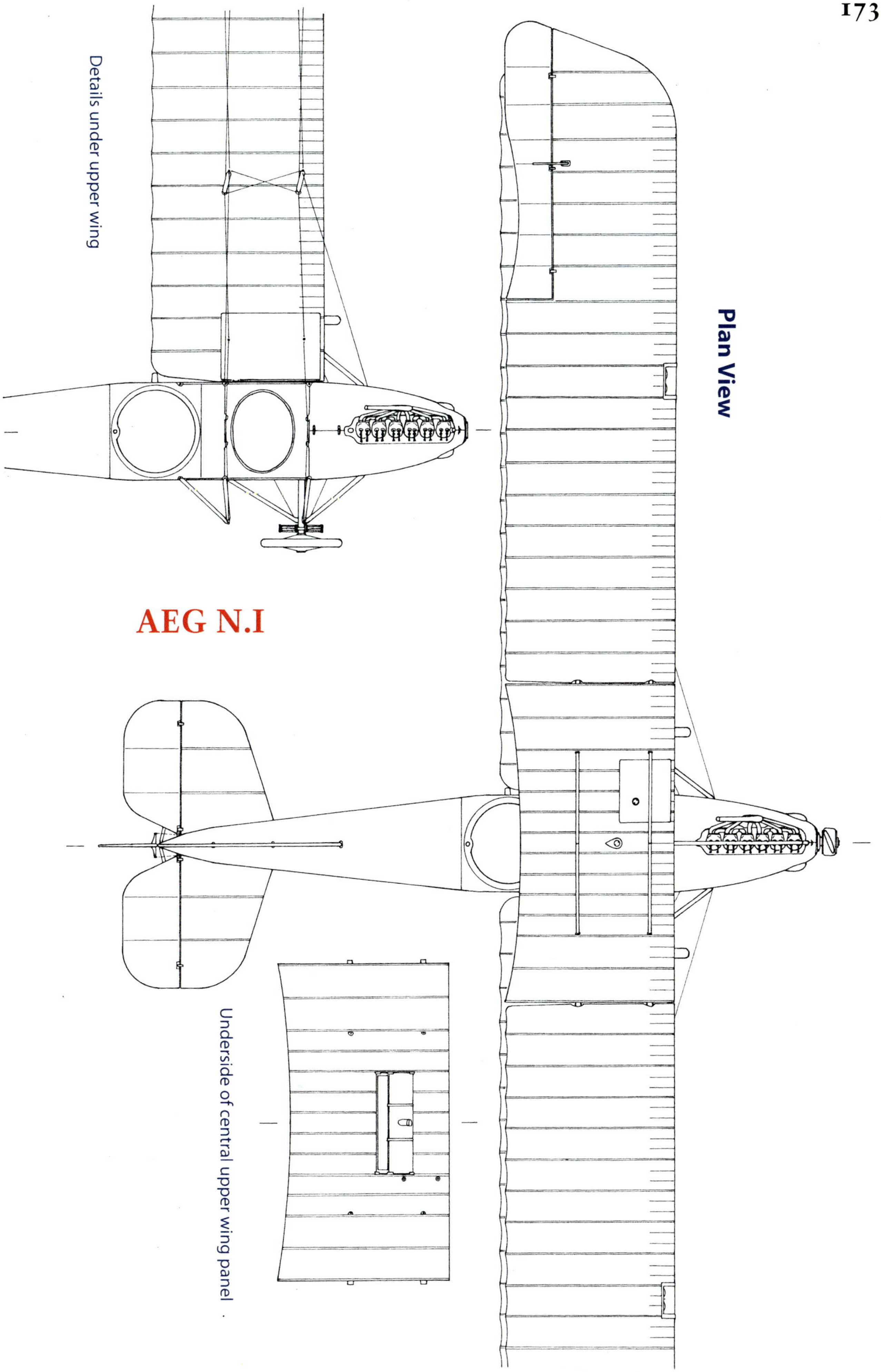
Details under upper wing
Plan View
Underside of central upper wing panel

AEG D.I

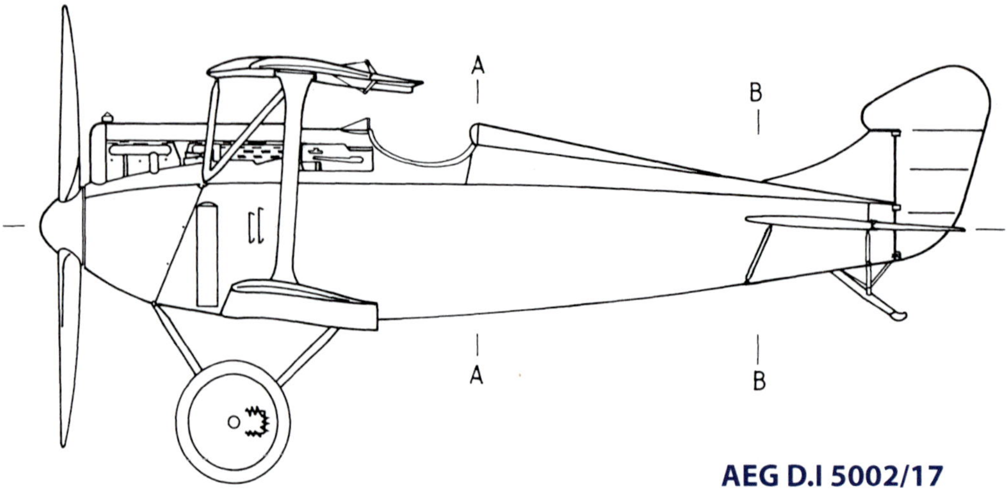

AEG D.I 5002/17

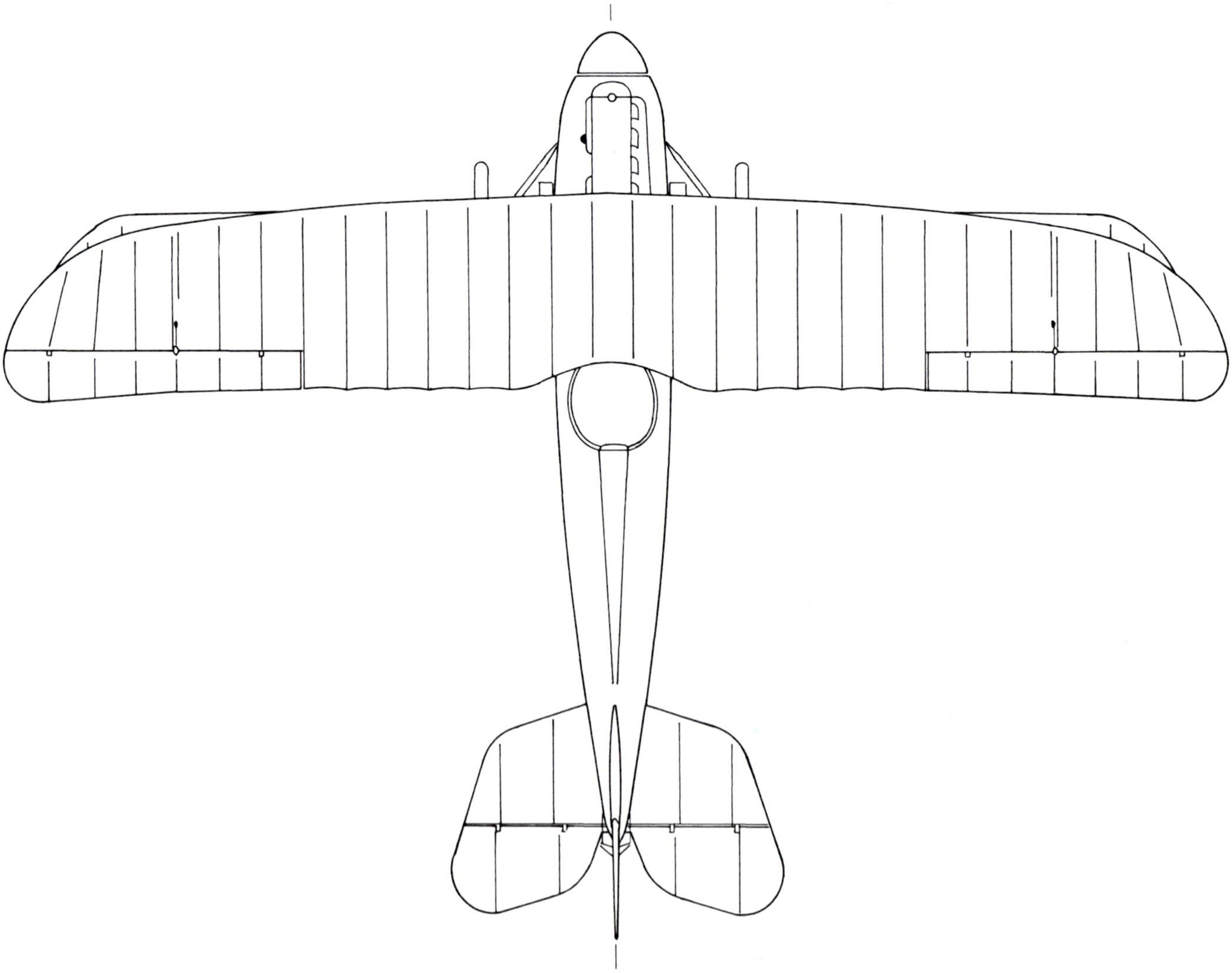

AEG D.I

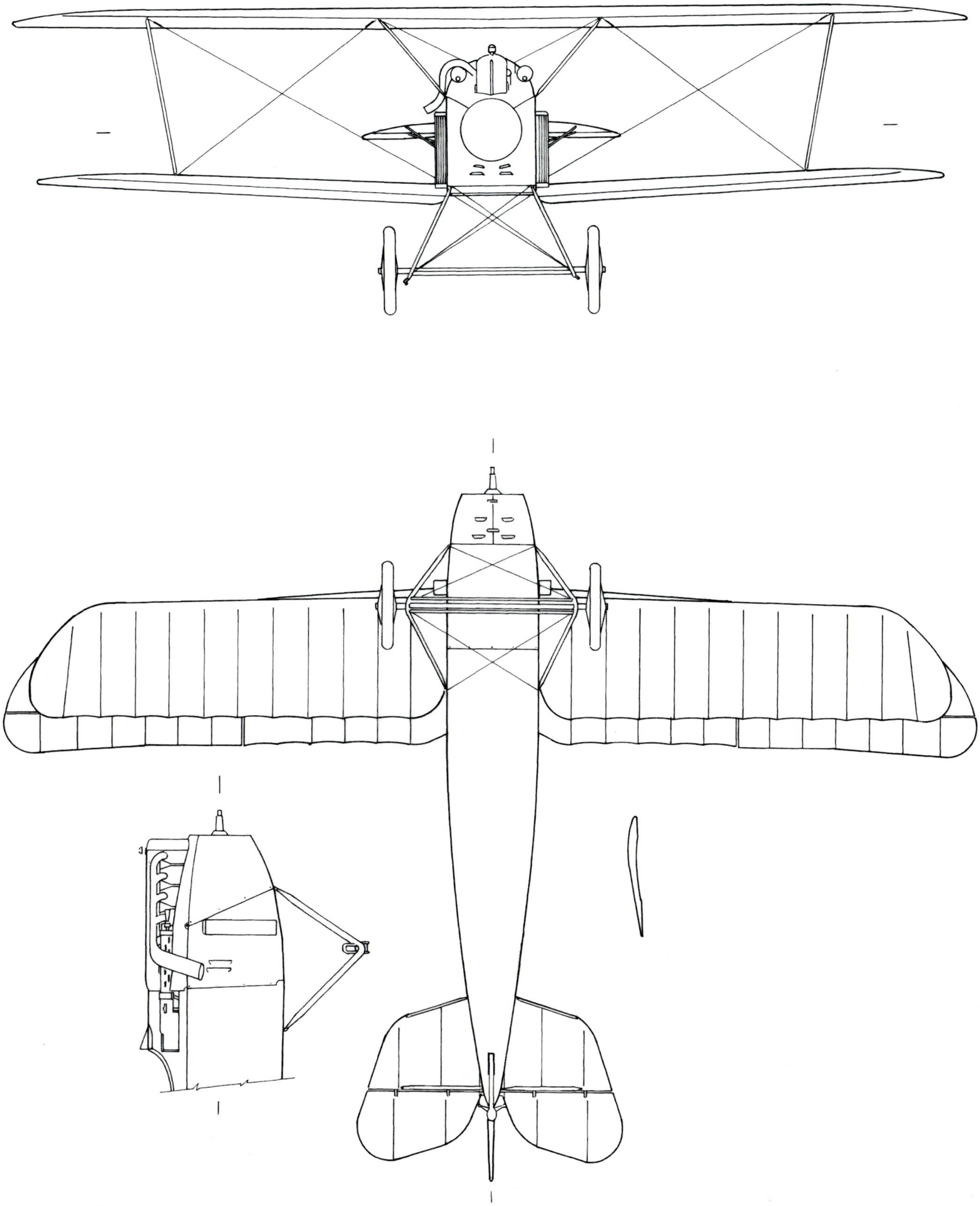

AEG G.II

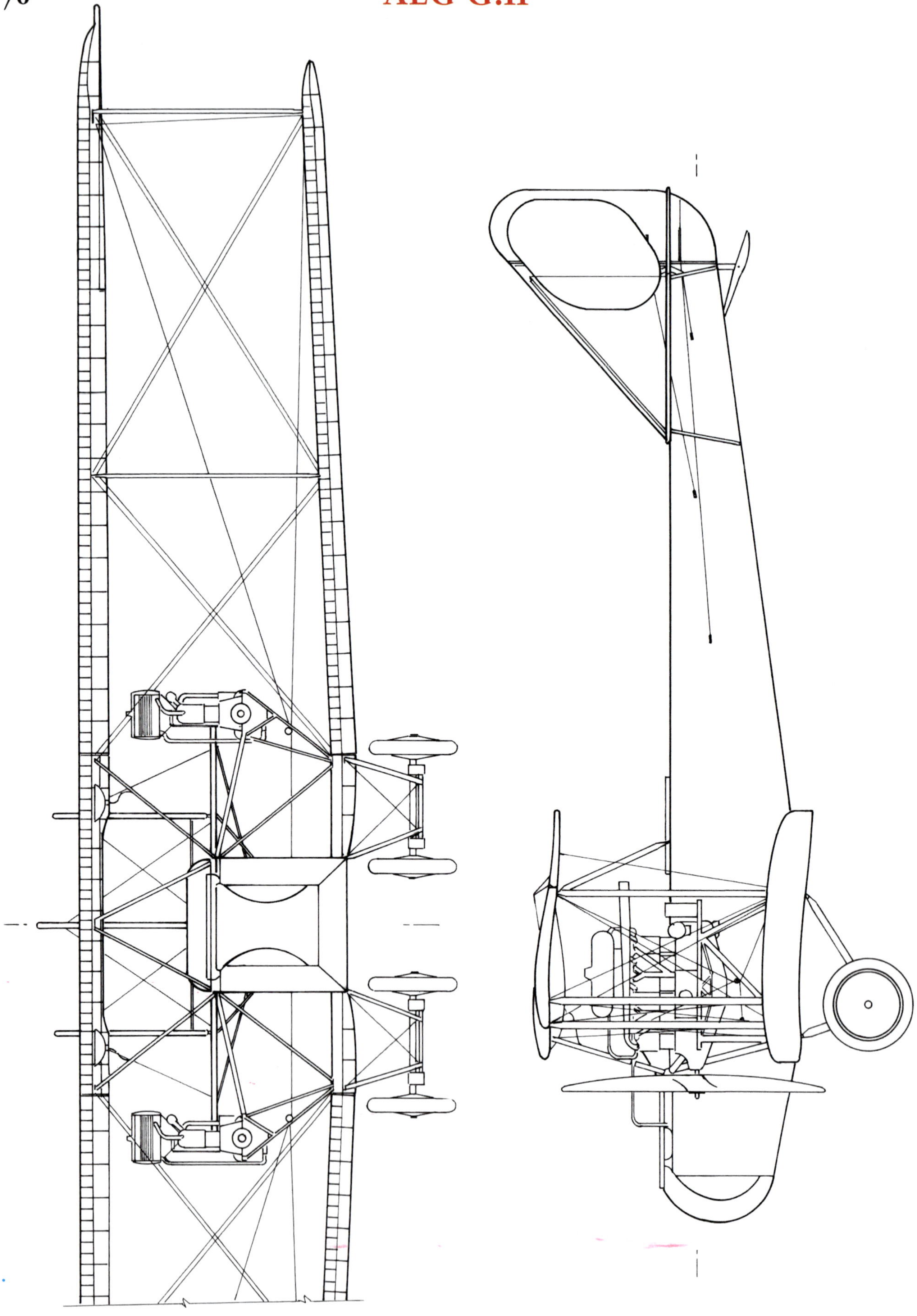

AEG G.II

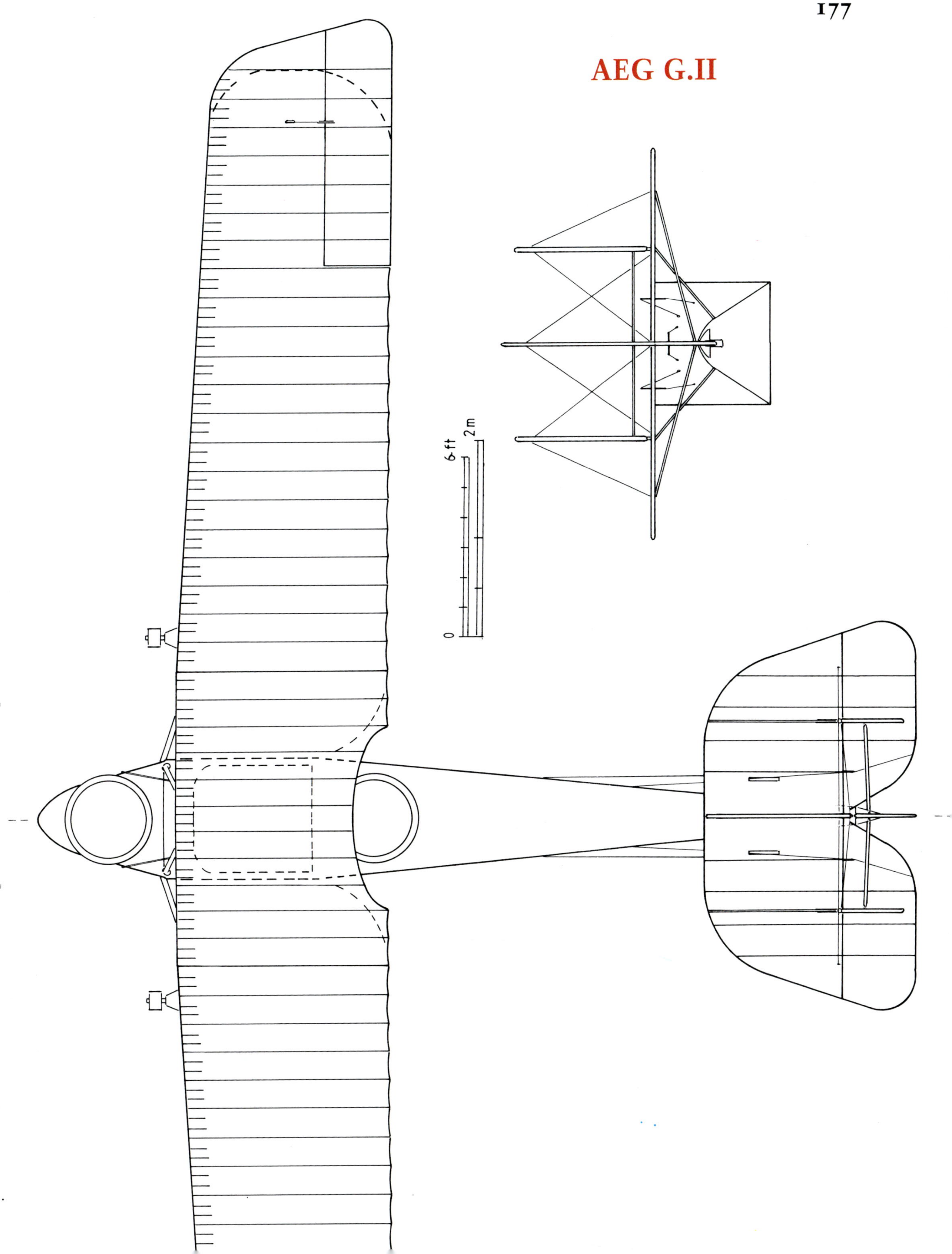

AEG G.III

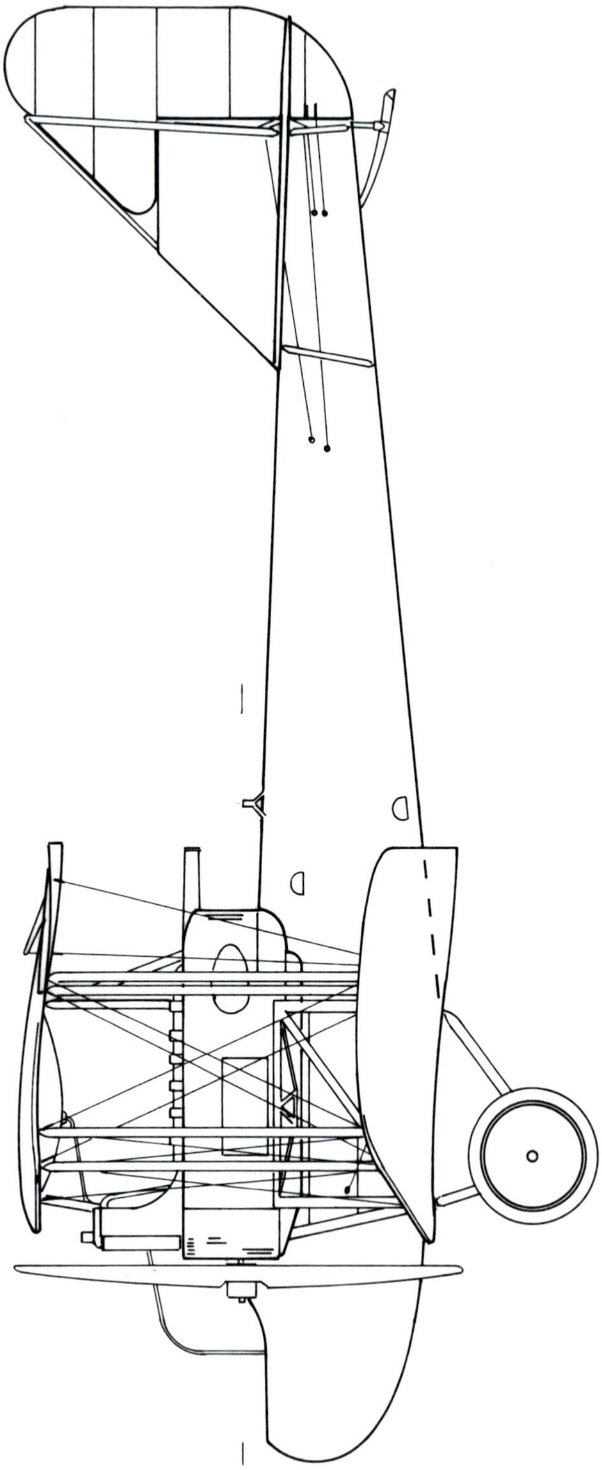

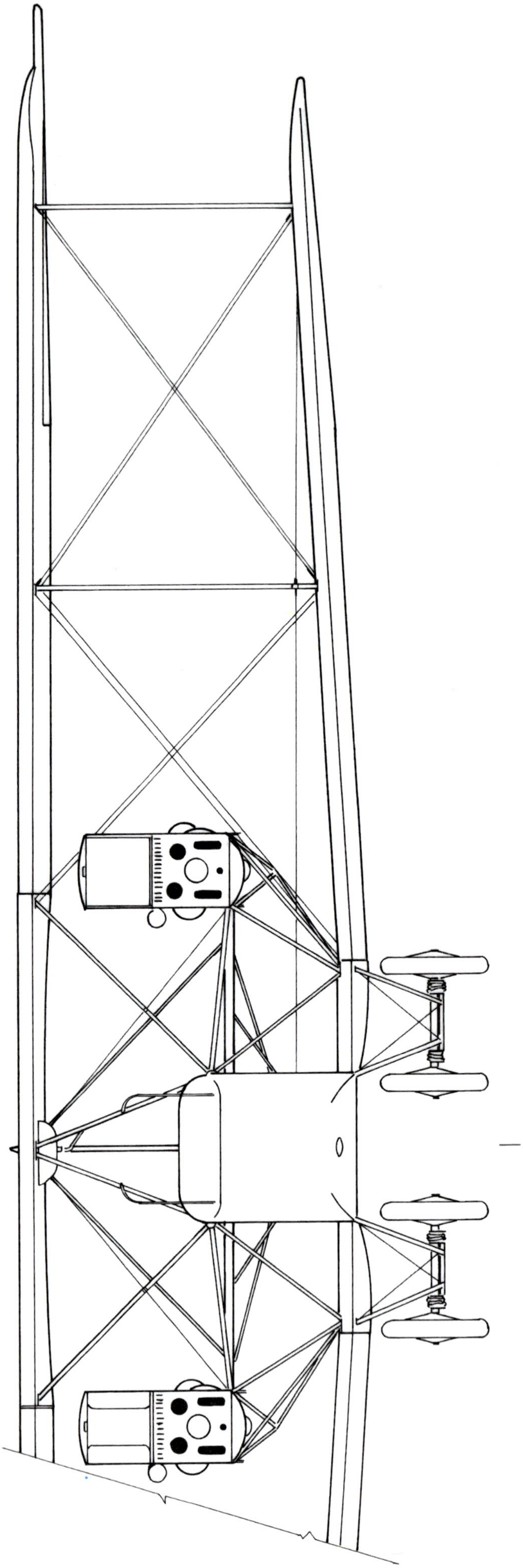

AEG G.III

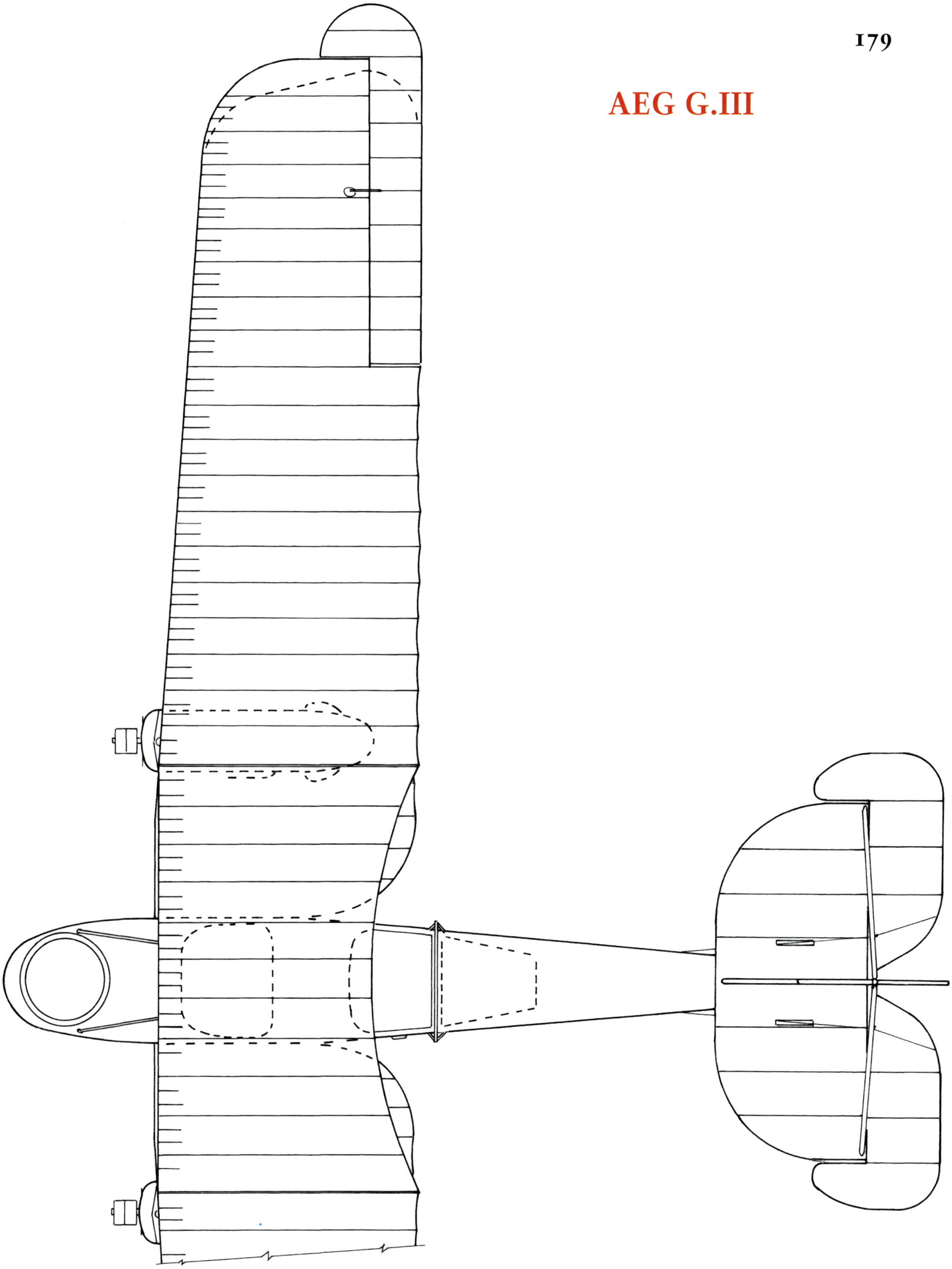

AEG G.IVk

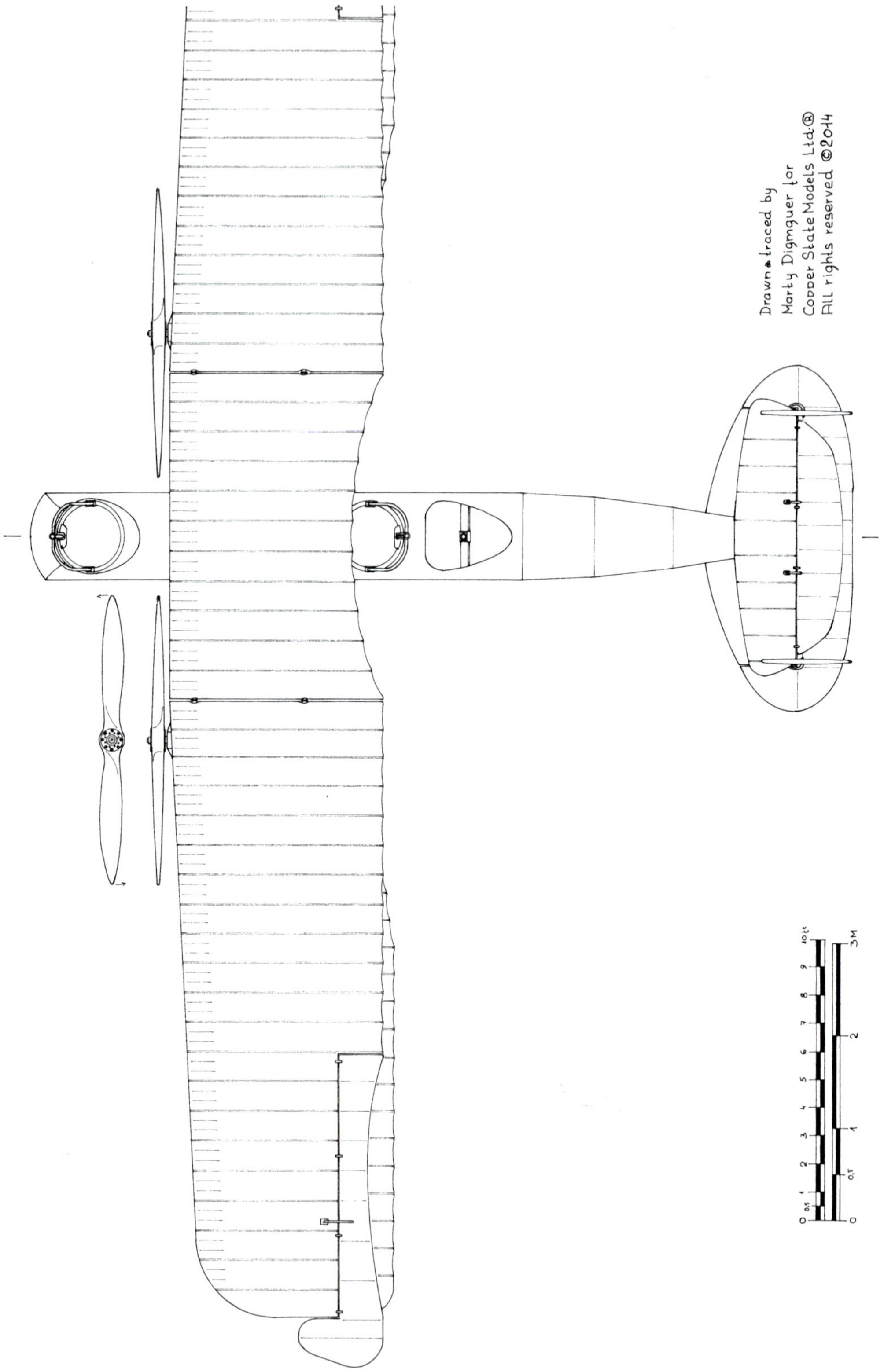
Drawn & traced by
Marty Digmquer for
Copper State Models Ltd.®
All rights reserved ©2014

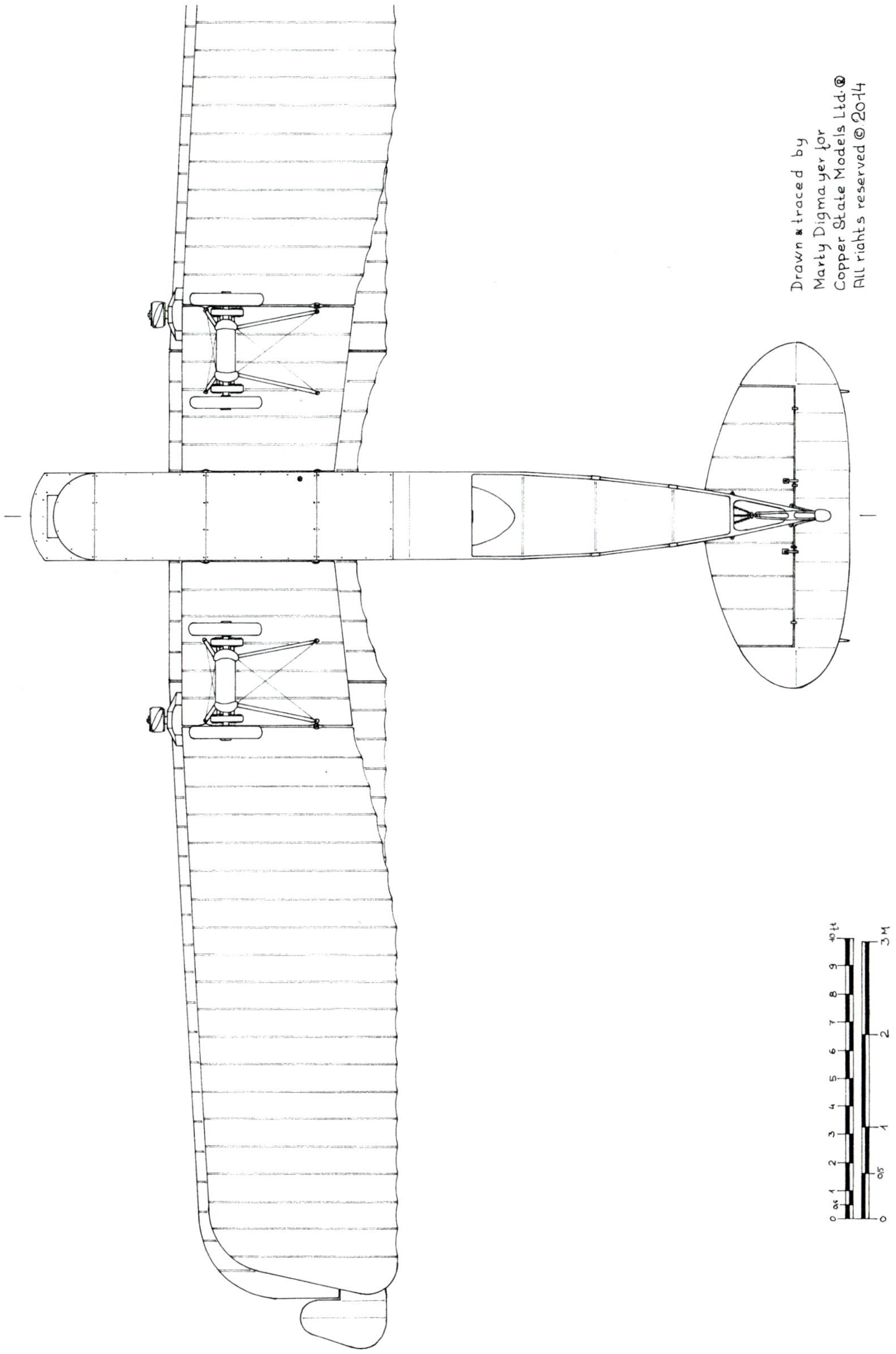
Drawn & traced by
Marty Digmayer for
Copper State Models Ltd.®
All rights reserved © 2014

AEG G.IVk

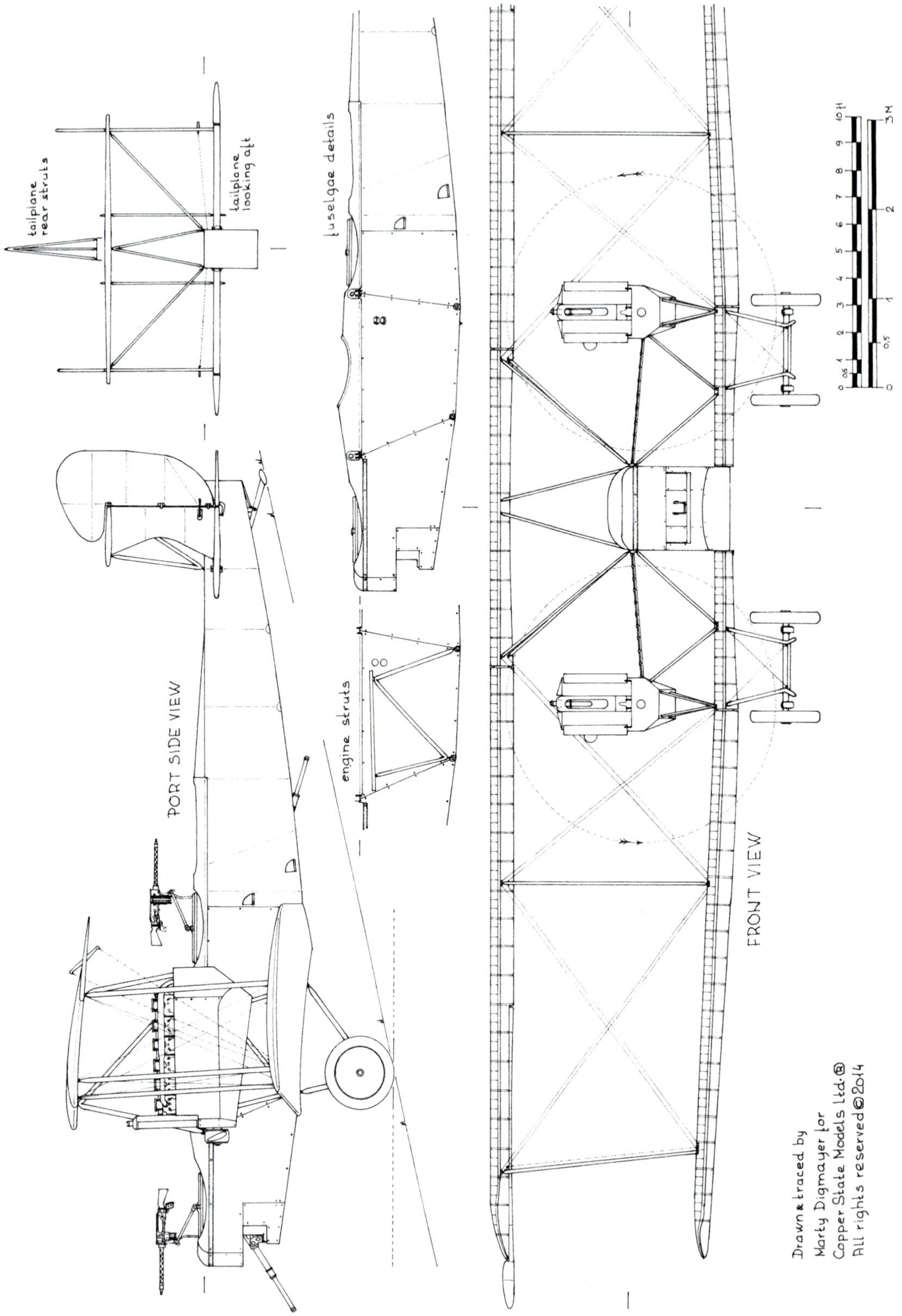
tailplane rear struts
tailplane looking aft
fuselgae details
engine struts
PORT SIDE VIEW
FRONT VIEW
0 0.5 1 2 3 4 5 6 7 8 9 10 ft
0 0.5 1 2 3M
Drawn & traced by
Marty Digmayer for
Copper State Models Ltd.®
All rights reserved©2014

AEG G.V

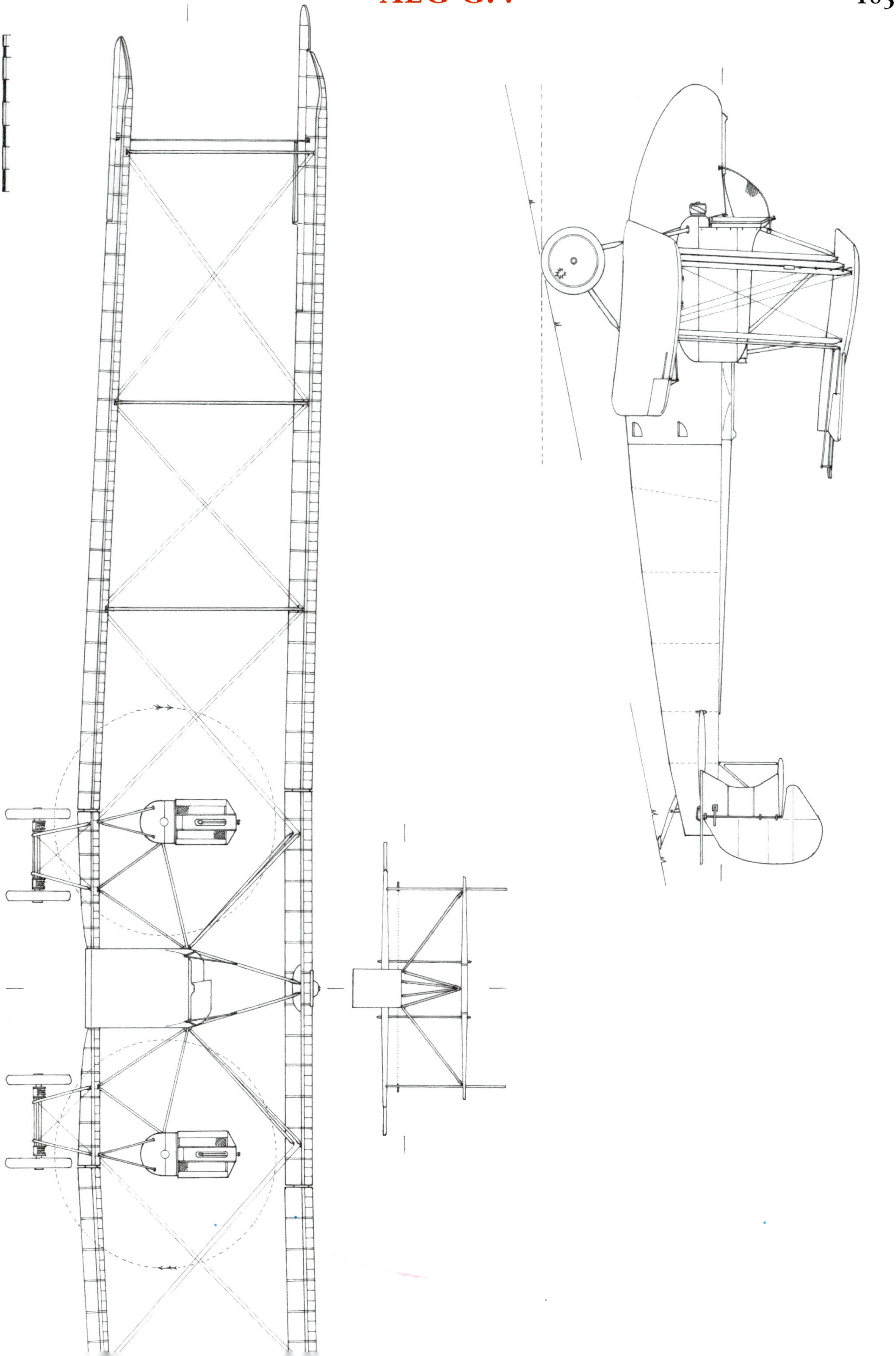

AEG G.V

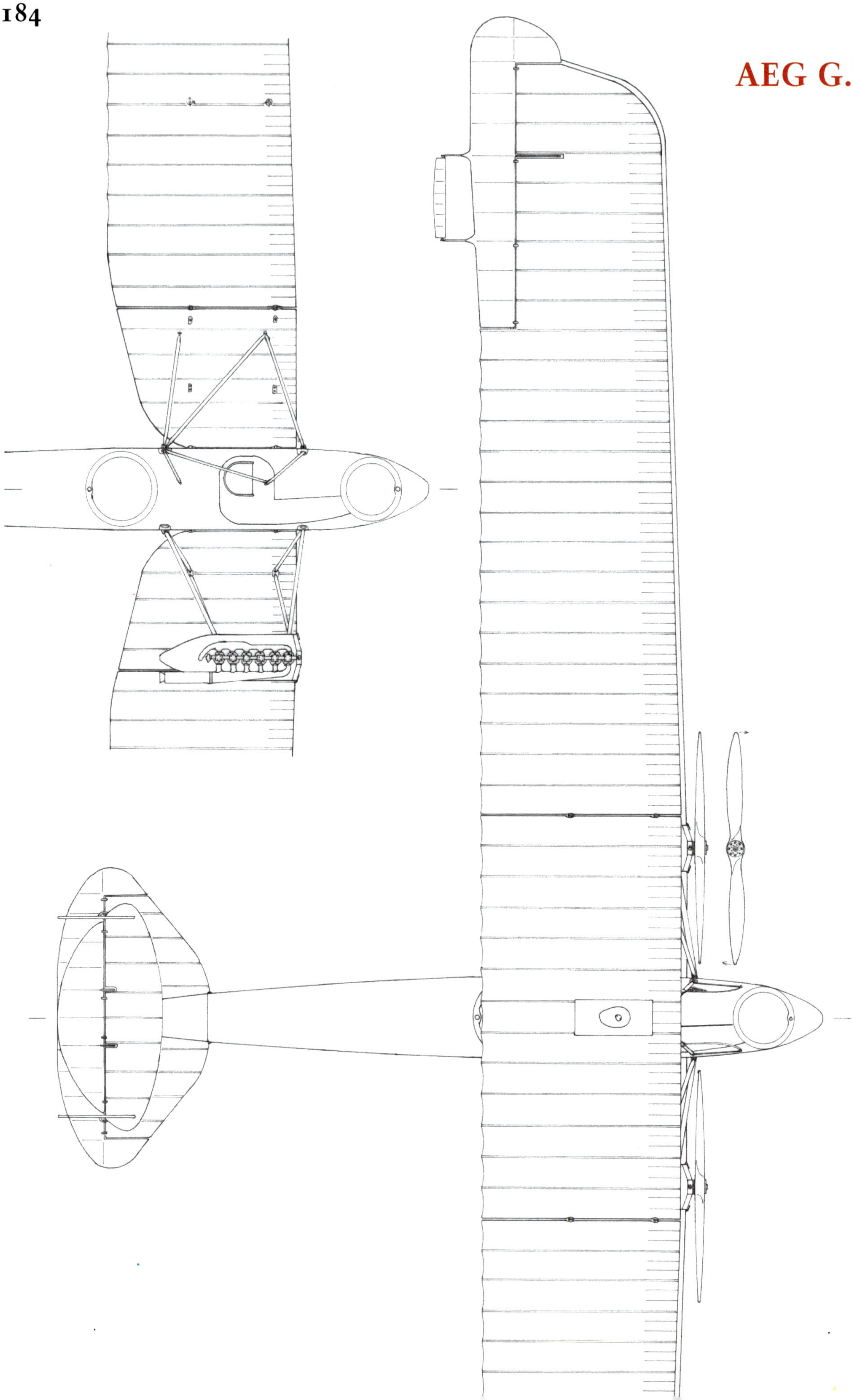

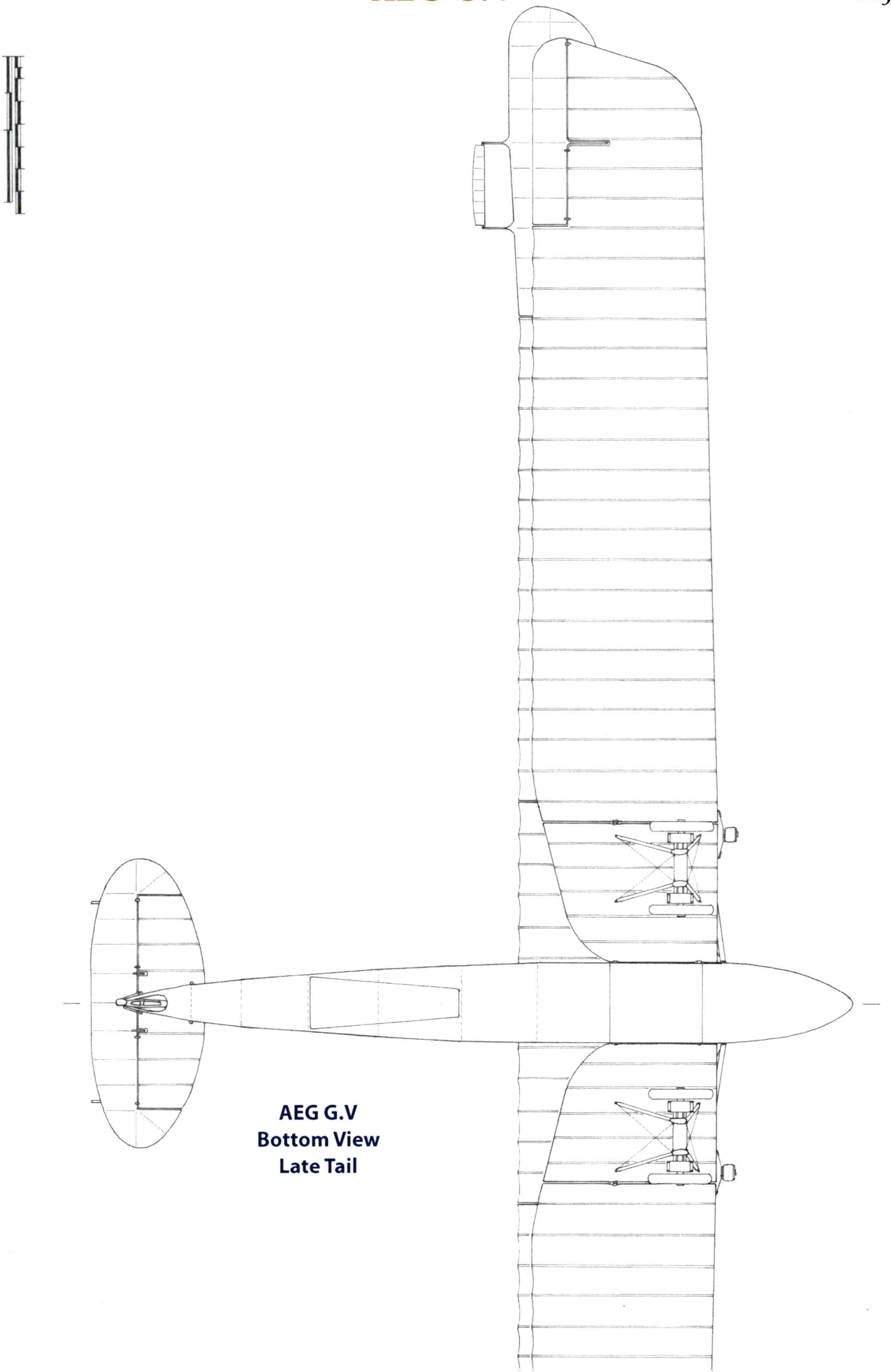

AEG G.V
Bottom View
Late Tail

Afterword

Although only one AEG fighter reached the front, and that had a tragic ending, AEG aircraft were in combat throughout the war and one AEG crewman was awarded the *Pour le Mérite* for his service while flying AEG bombers.

When war broke out Hermann Köhl was a 26-year old *Leutnant* in the *Württembergisches Pionier-Batallon Nr.* 13. After being wounded in action Köhl volunteered for the air service when he was deemed unfit for immediate return to his unit.

Köhl became an observer and joined *FA* 41. A year later he transferred to *Kampfstaffel* 22 of *KG* 4 flying escort and bombing missions. Köhl was aggressive and effective; on a night raid on Nov. 3, 1916, he and his crew blew up a huge Allied ammunition dump at Cerisy that burned out of control for three days. He received a number of awards and was wounded again. After recovery he learned to fly but remained an observer during the war. He also was assigned to lead *Kampfstaffel* 19 of *KG* 4. Köhl led the unit during its service in Italy and after it returned to the Western Front, he was given command of *BG* 7. On May 9, 1918 Köhl personally led the successful night bombing raid on an important ammunition dump at Blargies that destroyed it.

The next day, Köhl's name was submitted to the Prussian Awards Commission for the *Pour le Mérite*. His award was approved May 21; that same evening he was shot down during an exploratory night mission over Paris. Köhl survived but was captured. At the time he was not aware of being awarded the *Pour le Mérite*.

After the war many accomplished aviators were not immediately released and Köhl finally had to escape to Switzerland in September 1919.

Above: *Hptm.* Hermann Köhl wearing his *Pour le Mérite* and other decorations. The photo was taken post-war after Köhl escaped from captivity; he was captured after being shot down near Paris in his AEG G.IV.

Left: Fourteen aircrew of *Kampfstaffel* 22 of *KG* 4 pose with their 100 kg bombs in front of one of the unit's AEG G.IV bombers.

Made in United States
North Haven, CT
04 November 2026

21729702R00104